SOUND TECHNOLOGY AND THE AMERICAN CINEMA

Perception, Representation, Modernity

Film and Culture
John Belton, General Editor

FILM AND CULTURE

A series of Columbia University Press
Edited by John Belton

SOUND TECHNOLOGY AND THE AMERICAN CINEMA
Perception, Representation, Modernity

James Lastra

COLUMBIA UNIVERSITY PRESS
NEW YORK

Columbia University Press
Publishers Since 1893
New York   Chichester, West Sussex
Copyright © 2000 by Columbia University Press

Library of Congress Cataloging-in-Publication Data
Lastra, James.
Sound technology and the American cinema : perception, representation, modernity /
James Lastra.
p. cm. — (Film and culture)
Includes bibliographical references and index.
ISBN 0–231–11516–4 (cloth : alk. paper) — ISBN 0–231–11517–2 (paper : alk. paper)
1. Sound motion pictures—History. 2. Sound—Recording and reproducing—History.
I. Title. II. Series

PN1995.7.L37 2000
791.43'09—dc21   99–087737

Casebound editions of Columbia University Press books
are printed on permanent and durable acid-free paper.
Designed by Chang Jae Lee
Printed in the United States of America
c 10 9 8 7 6 5 4 3 2 1
p 10 9 8 7 6 5 4 3 2 1

For Janice, Charlotte, and Paul

CONTENTS

ACKNOWLEDGMENTS

Like most projects, this book could not have been accomplished without the aid of many others in matters both large and small. From its first hesitant drafts in 1989, until its completion ten years later, this book has benefited from the wise counsel offered by friends, colleagues, and teachers in several different places and institutions. The earliest versions of the ideas set forth here were written at the University of Iowa. My debt to teachers and fellow students there is enormous. Professors Dudley Andrew, John Peters, Lauren Rabinovitz, and Steven Ungar all made valuable and irreplaceable contributions to its early growth, as did fellow students Charles O'Brien, Steve Wurtzler, Dana Benelli, Scott Curtis, Pieter Pereboom, Greg Easley, and James McLaughlin. I am grateful for the many enlightening conversations I have had during my many return trips to Iowa City since graduation, especially those with the members of the Sound Research Seminar.

My greatest Iowa debt, however, is to Rick Altman, without whose camaraderie, insight, and constant intellectual challenges neither the dissertation nor the book would have been written. Rick's influence is evident on every page, and it would be a far poorer book were it not for his inspiration, his criticism, and, most of all, his friendship.

The Academy of Motion Picture Arts and Sciences and the Warner Bros. Archives at the University of Southern California aided my research immeasurably. I would especially like to thank Valentin Almendarez, Barbara Hall, Scott Curtis, and Stuart Eng for their kind, and usually crucial, assistance. The staffs at the Chicago Public Library and the Special Collections Departments of the Joseph Regenstein and John Crerar Libraries at the University of Chicago were unfailingly helpful, in spite of the frequent obscurity of my requests.

Don Crafton, Mary Ann Doane, Martin Jay, Tom Levin, Terry Smith, Chuck Wolfe, and Ed Branigan deserve special thanks for reading and listening to all or parts of this book and for making significant criticisms, suggestions, and contributions over the past few years. At Columbia University Press, I have had the privilege to work with John Belton, Jennifer Crewe, and Roy Thomas, who have been unfailingly patient, even when disasters repeatedly hit.

The intellectual generosity of my colleagues at the University of Chicago has been remarkable and touching. Tom Gunning, simply put, changed the nature of the project through his comments, encouragement, and intellectual guidance. Jim Chandler, Tom Mitchell, Katie Trumpener, and Bill Brown provided invaluable help with particularly knotty issues. Josh Scodel and Lisa Ruddick offered sound advice on structure and rhetoric. Students too numerous to mention served as sounding boards for ideas not yet committed to print, and hammered them into shape in class discussion.

My deepest affection and respect go to Miriam Hansen, who has done more to shape my life and work over the past few years than I can adequately express. Intellectually fearless, committed, brilliant, and original, she has served as a model for what a scholar and a teacher should be. I have learned form her not only how to be a better academic but also how to be a better colleague. More important, she has shown me what it means to have strength, integrity, and wisdom. The book spends a great deal of time pondering the inhuman. Miriam has expanded my understanding of what it means to be truly human.

Finally, I would like to thank my family for their unflagging love and support over the past few, difficult, years. My mother, Jean, and sister, Joan, helped in thousands of small but essential ways, and encouraged me at all the right times. My wife, Janice, and my children, Charlotte and Paul, above all, have given purpose to the entire enterprise, and joy to my every day. I cannot thank them enough. I dedicate this book to them and to the memory of James F. Lastra, James McLaughlin, and Michael Altman.

SOUND TECHNOLOGY AND THE AMERICAN CINEMA

INTRODUCTION: DISCOURSE/DEVICE/PRACTICE/INSTITUTION
Representational Technologies and American Culture

> *In the photographic camera [man] has created an instrument which retains the fleeting visual impressions, just as the gramophone disc retains the equally fleeting auditory ones; both are at bottom materializations of the power he possesses of recollection, his memory. With the help of the telephone he can hear at distances which would be respected as unattainable even in a fairy tale. . . . Long ago he formed an ideal conception of omnipotence and omniscience which he embodied in his gods. . . . Man has, as it were, become a kind of prosthetic God. When he puts on all his auxiliary organs he is truly magnificent; but those organs have not grown on to him and they still give him much trouble at times.*
> —Sigmund Freud, *Civilization and Its Discontents*

Observing passersby from the window of a London coffee shop, the narrator of Edgar Allan Poe's "The Man of the Crowd" (1840) becomes so emotionally and physically distressed that he rouses himself from his seat and pursues for twenty-four hours a face that "like a certain German book . . . does not permit itself to be read." In a story that has been described, alternatively, as an "x-ray" or "the embryo" of the detective story, Poe presents us with what one scholar has called a specifically modern and urban crisis of "legibility."[1] Confident in his capacity to "read" the types in the crowd ("the tribe of clerks," "the race of swell pickpockets," "the gamblers," "the Jew Peddlers"), our narrator is brought up short by a face whose "absolute idiosyncrasy" remains, despite great effort, illegible.

Undoubtedly, as Walter Benjamin and others have maintained, "The Man of the Crowd" registers anxieties brought about by modernization and the growth of the city, ranging from increased fear of crime, to loneliness, alienation among a sea of strangers, loss of tradition, a relentless assault on the senses, and the threat of catastrophic, industrial accident. Poe's detached observer rests complacently on his perch until his visual mastery is threatened by the appearance of a singularity that finds no place within the grids of intelligibility through which he customarily domesticates the city's diversity. Frustrated, the narrator can only explain the illegible as a symptom of "the genius of deep crime." Indeed, as Dana Brand suggests, illegibility, in a sense, *is* the crime. Yet the threat represented by either the city or the unknowable stranger does not just respond to the contingencies of modern experience. Undoubtedly, "The Man of the Crowd" registers these things—it *is* a story about urban modernity—but it can also be read more specifically—and provocatively—as a story about the challenges to sensory experience raised by photography.

From this point of view, it is hardly insignificant that Poe's story hinges on the problems posed by singularity to habitual modes of perception and experience, nor that its entire mise-en-scène serves to figure this transformation as photographic. From the detached but highly sensitive observation of a narrator situated in a darkened room behind a mediating layer of glass that projects, lenslike, from the front of the building, to the perceptual dialectic of familiarity and pure idiosyncrasy, the story offers an allegorical staging of the epistemological drama of technologically mediated sensory experience, where habitual modes of looking and knowing confront their limitations in a disorienting encounter with the contingent real. The man of the crowd, in effect, *is* a photograph—or the world *as* photographed—simultaneously compelling and threatening, signaling the fragility of our familiar ways of knowing, while trumpeting the arrival of new and disconcerting epistemologies.

Joining forces with the enormously popular "anthropological" guides to 1840s Parisian social types—the *physiologies*—and the classifications deployed by our coffee drinker, photography soon emerged as an important tool for classifying, rationalizing, and finally mastering, the visible world.[2] Within a range of practices defined on the one hand by eugenicist Francis Galton's attempts to define ethnic and social types by way of "composite portraits" made by superimposing twenty or thirty photographic portraits of "the Jew," "the Pole," "the Irishman," or "the criminal" and, on the other, by Alphonse Bertillon's widely adopted photographic system for identifying criminals, our narrator's perceptual and ideological stereotypes seem less than innocent.[3] Nor can it seem merely ironic that Poe's narrator can only "read" the illegible as evidence of crime, for photography was soon to become the primary representational form through which the state identified, classified, and rendered knowable the vast, anonymous populations characteristic of the city.

However, in spite of photography's role in making the visible world more and more legible through a variety of rationalizing pictorial practices, it simultaneously exposed modernity's underside, its randomness, arbitrariness, and irrationality. The photographic encounter with the world, we might say, is always a gamble of sorts through which traditional representational forms like the portrait open themselves structurally to the aleatory, the idiosyncratic, the unintelligible.[4] Photography's inhuman tendency to wrest the completely singular and unexpected from that encounter can result in an image as utterly unintelligible as the face in Poe's story. The appearance of the unnatural-seeming photograph heralded an era in picturing where the ephemeral and contingent assumed an unprecedented parity with the stable, the universal, and the eternal; and while the photograph's essential contingency can hardly register everything significant about the history of modernity, it directs us toward a segment of that history, which, while partial, is nevertheless crucial.

But neither the story nor modernity itself is entirely defined by the visual. Despite Poe's relentlessly optical obsession—every paragraph evokes the visual at least once—the aural murmurs in the background as the text's and modernity's repressed sense. While the story's master trope—the unreadable book—implies a purely visual barrier to knowing the Man and his secret, Poe elaborates the unreadable as the *untold*, the never-confessed. And regardless of Poe's tendency to imply that his observer is engaged in the purely silent contemplation of the crowd, the city's voices creep in again and again, in the gambler's "guarded lowness of tone in conversation," the "inarticulate drunkards," the organ-grinders, the "ballad mongers," and in the "ragged artizans and exhausted laborers of every description . . . all full of a noisy and inordinate vivacity which jarred discordantly upon the ear, and gave an aching sensation to the eye."

Even the narrator's fevered pursuit of the stranger seems haunted by the sounds of the city, since he remains unable to imagine that the man who so studiously avoids noticing him may simply have *heard* him following on London's cobbled streets. While the narrator finds it prudent to "walk close at his elbow for fear of *losing sight of* him," the stranger seems to rely less on vision. "Never once turning his head to look back, he did not observe me," says the narrator, unaware of the role that hearing, smell, and touch might play in the urban epistemology of the Man of the Crowd, or how they might form an alternate economy of the senses. But neither the narrator nor Poe is alone in his refusal to consider the aural dimension of modernity, since generations of critics have just as thoroughly avoided it in their considerations of the story and of modernity itself.

If we listen a bit more attentively, though, the nineteenth century tells a quiet but no less powerful story of aural modernity and modernization. Vision was neither the only sense to be transformed nor the only one to act as an agent of transformation. Hearing was just as surely dislocated, "mobilized," restructured, and mechanized. The "annihilations" of space and time affected hearing as much as seeing, and acoustic experience was as thoroughly commodified as its optical counterpart. Vision and hearing are the senses that, as a consequence of nineteenth-century innovations, have been most fully penetrated by technology and that have, reciprocally, shaped sensory technologies to the greatest degree.

Regardless of its myriad other causes, the experience we describe as "modernity"—an experience of profound temporal and spatial displacements, of often accelerated and diversified shocks, of new modes of sociality and of experience—has been shaped decisively by the technological media. The cinema above all has come to stand for "modernity" itself, seeming to emblematize in the most compelling and even visceral way, the frequently violent shifts in social and cultural life, especially the newly possible (if not inevitable) forms of spatial, temporal, and sensual restructuring.

The violent social and physical changes wrought by capitalist industrialization, the telegraphic, telephonic, phonographic, and photographic "annihilations" of time and space, the utopian yet threatening role played by these technologies as prosthetic sensory organs, and the new modes of mass production, distribution, and consumption—these are nowhere better concretized than in the movie theater, where sounds and sights from disparate times and from all parts of the world come together in an assembly-line-like progression of lightning perceptions destined for an anonymous but statistically describable audience. The institutional deployment and industrial exploitation of cinematic technology ramifies throughout the culture we call modern, shaping our experience of others, of history, of ourselves. Still, cinema, the most pervasive mechanism for disseminating technologically mediated sensory experience is, and was from its very inception, not a *visual* phenomenon but a resolutely *audio*visual one.

This book, therefore, attempts to address this fact by combining a history of modernity with a study of cinema sound. Bringing together studies of modernity and studies of recorded sound may not appear obviously necessary, but both enterprises stand to profit immensely from the encounter. I argue two basic claims from this intersection: first, that aurality has been the unthought in accounts of modernity and that, consequently, we have overestimated the hegemony of the visual; and second, that modernity has been underexamined in accounts of recorded sound, and cinema sound in particular. In short, by working in a mode that might be called "thick epistemology," I try to bring "sound" into modernity, and modernity into "sound." To address sound film technology from a historical perspective, in fact, necessarily entails this process since the cinema is as unthinkable outside of modernity as modernity is unimaginable without the cinema.

What "modernity" and "sound" share is therefore a central concern of this book. However, rather than attempt to account for modernity in its fullness and complexity or for sound in the abstract, I use their intersection in a specific conjunction as a way of framing both. As my allegorical reading of Poe suggests, I am principally concerned with a particular "slice" of modernity and the myriad reflexive attempts to theorize it over the past 160 years. I am particularly interested in the problems posed by the technological recording of sensory data to aesthetic experience both in its broad, etymological sense as the science of perception and in its more narrow and familiar sense as the study of the institutions of art. When modern technologies made it suddenly possible to record and reproduce images and sounds without the intervention of a human subject, the problems of contingency, chance, and arbitrariness thrust themselves into the realms of perception and of aesthetic production with equal force. This dual crisis defines the historical background both for the emergence of these media as representational forms in general and for the coming of sound to Hollywood in particular. It is my contention, in fact, that twentieth-century filmmakers and sound technicians persistently restage nineteenth-century encounters with technological modernity in an attempt to come to grips with the problems posed to definitions of the human by what we might call the mechanical senses.

As Miriam Hansen has argued, it simply will not do to leave modernity in the nineteenth century and, what is more, in Baudelaire's Paris.[5] "Modernity" is still very much with us, and both the cinema and modern life achieve a certain maturity and scope only in our own century. Since twentieth-century cultures, and Hollywood in particular, have repeatedly engaged in fraught confrontations with new representational technologies—ones that inevitably restage and reenact earlier moments—it is neither possible nor desirable to understand our own engagements in ignorance of prior ones. Technology's role in the processes of modernization, the aesthetics of modernism, and the experience of modernity, especially as it was negotiated in the nineteenth century, established the terms through which we still (in refracted and transformed ways) confront, describe, and assimilate new representational forms and possibilities.

Although it is important to avoid simply repeating the pieties of research about the relationship between media and modernity—the decisive and hegemonic rise of the visual, the annihilation of space and time, the utter transformation of perception, and so on—it is likewise important to remember and re-ask the questions that prompted these familiar claims. The most compelling and incisive investigation of the relationship between perception, representation, and modernity has been produced by those in and around the Frankfurt School (Siegfried Kracauer, Theodor W. Adorno, and Walter Benjamin), and I will re-ask many of their questions using new perspectives and new archives.

Benjamin's analysis of modernity is invaluable for illustrating that experience itself, especially sensory experience, can be the object of sustained historical and political analysis. And in this book the history and politics of the technological mediation of experience are ultimately at stake. So deeply do the technical media reach into our basic understandings of the nature of experience that each emergence—the daguerreotype in 1839, the phonograph in 1877, motion pictures in 1893–1895—occasioned far-ranging debates about the very nature of humanity and human experience. That these media's apparent extension, perfection, replacement, or destruction of human faculties had a profoundly transformative effect is witnessed by their persistent analysis as prosthetic sensory apparatuses. From Étienne-Jules Marey, to Thomas Edison, to Sigmund Freud, to Benjamin, to Marshall McLuhan, to Susan Buck-Morss, critics have described the new media as, in some important sense, "prosthetic." It is hardly surprising, then, to learn that the phonograph was initially classified with and discussed in terms of speaking automata, that its earliest incarnations were often constructed from dissected ears, or that its scientific counterparts came to stand metaphorically for the ideal attitude of the human researcher.

In a more psychological and political vein, Benjamin, Freud, and Buck-Morss discuss the camera and phonograph as principal components of a new kind of protective sensory apparatus—a "second skin" as it were—that each describes as a "stimulus shield." This "anaesthetizing" layer, they argue, emerged as a crucial component of a modern consciousness continually assaulted by experiential "shocks." As both a source of those shocks, and perhaps an antidote to them, the technical media assumed a more directly political function, as one of the prime arbiters of experience. On the one hand, they threatened to intensify and expand the numbing of consciousness Buck-Morss explores, leading to what Benjamin analyzes as the fascist "aestheticization of politics" and, on the other, offered the possibility of functioning as a producer of the images or experiences through which a collective might come to recognize itself and its own material conditions of existence. By coming to grips with the technologically mediated conditions of modern sensory experience, by "working through" these technologies rather than dismissing them, Benjamin hoped the utopian potential of the technical media might be redeemed.

Along a different axis, signaled by the ubiquitous suffix "graph," these same media were understood through their complex relationship to writing. Scriptural tropes emerged to describe the impact and significance of the new media. Whether it be Edison's or Alexander Graham Bell's invocation of hieroglyphics and universal languages, or the former's almost Platonic analysis of the phonogram's ability to preserve, extend, and distance speech, writing seemed to offer a figure through which to describe some of the more disconcerting and enabling implications of the new media. What is more, by preserving the previously ephemeral in a fixed and repeatable form, these media replaced writing as the premier historical storage technology. Indeed, it is not unusual to find all three media discussed primarily as a kind of mnemotechnics, solidifying their metaphorical connections to writing and inscription.

Whether troped as automata or as forms of inscription, the new media raised astonishingly complex epistemological issues. Here, suddenly, were images executed by *machines*—perhaps even accidentally—where human activity entered only to control otherwise autonomous mechanical processes. Here, in effect, was a whole new class of signs whose referential status was not only confusing but critically so. Why was photography generally understood as a more accurate visual record than any other kind of picture? When did people learn to attribute certain formal features, like extreme points of view or blurriness, to processes of inscription? How, in short, did people learn to deal, in both a discursive and practical manner, with these new phenomena? How did the figures of human simulation and of inscription mediate these processes, and how are they imbedded in modernity?

Each trope for understanding and normalizing the new media—as human prosthesis or as a form of writing—in fact describes a different facet of the same threat: the rising specter of the inhuman within human experience posed by mechanical inscription in general. While industrial capitalism dehumanized laborers in the workplace, the new media threatened to do the same in other ways in other venues. Both writing, where signification could occur contrary to human intention or even presence, and simulation, where a kind of Frankenstein's monster could usurp and surpass bodily sense organs, threatened to dwarf human capacity, outdate it, render it obsolete, and displace it from its traditional (and "rightful") home altogether. Metaphors of simulation and inscription, however, responded to those challenges, working to neutralize them, render them legible, suggest techniques of conceptual mastery, and even provoke avenues for instrument design and representational practice. By and large, these metaphors enable the inhuman aspects of technology and technological inscription to be reanthropomorphized, in the nineteenth and twentieth centuries alike. In short, these figures helped identify and announce cultural problem spots, but also helped domesticate them.

Hollywood's twentieth-century responses to new media have been governed by responses borrowed, as it were, from the nineteenth century. Understanding innovations in terms of either writing or human simulation (however different from their earlier counterparts) served again and again as discursive and conceptual tools for identifying, segregating, and controlling technical and representational possibilities. And just as it becomes easier to understand the coming of sound to Hollywood with the nineteenth-century reception of the phonograph as a background, it conversely becomes easier to grasp the significance of the latter from the perspective of the former.

Still, despite the rather broad questions that inform its outlook, the more specific goal of this book is to track the development of a new representational technology—sound recording—from its emergence to its confrontation with and integration into the Hollywood film industry. Given these ambitions, to look back to the late eighteenth century—to natural philosophy, stenography, automata, and human physiology—is a deliberately provocative act designed to produce a genealogy that links the present moment both to the recent past of the classical Hollywood cinema and to the more distant past of the emerging modern world. I forge these links by tracking the shifting relationships established between our senses, technology, and forms of representation, seeking to understand how the experiences of technological modernity shaped not only our attitudes toward mechanical recording and representation, but also how ideas about the senses shaped the construction of phonographs and cameras, and how those devices reciprocally shaped our understanding of the senses and sensory experience. Finally, the question is how our altered conceptions of both technology and perception then shaped our sense of aesthetics.

The material history of sound devices and the history of their practices are unthinkable without this genealogy. What may at first appear as a fanciful journey I construct from talking heads, to talking dogs, and from there to universal alphabets, laboratory experiments, and decapitated pets, emerged from a concrete and rather specific question: why did audio engineers of the 1920s inevitably assume that sound recording should always literally mimic the sensory experiences of an imaginary body in real space? Looking at the standard and widely available histories of sound recording, I noticed that every one of them made passing and apparently pointless reference to a succession of talking heads created in bygone eras. In unpacking those references, I came to realize that, even in the late 1920s, commentators and practitioners were still grappling with the basic questions of representational modernity—namely, how do we understand the relationships between perception, representation, and technology? That is, are photographs or recorded sounds best understood as real human perceptions, or as something quite other? Therefore, the goal of the first two chapters of this book is to reactivate the mute history that shapes the emergence of this century's most influential audiovisual form—the classical Hollywood sound film.

Since most of the relevant questions about the coming of sound to Hollywood cannot be asked of the nineteenth century, the remaining four chapters examine the period 1925–1934 in four different registers. Chapter 3 examines the emerging institution of the classical cinema through the norms and practices of "silent" film sound accompaniment, and thereby sets the stage for the sound film. Asking anew what it might mean, and what, in fact, it did mean for sound and images to "go together," I seek to defamiliarize our view of the recent past in order better to see its specific contours, and its unfamiliar potentials. Looking at the surprising variety of "synchronization" techniques, and rethinking their functions, I construct a complex field of practices and functions that form the horizon governing the adoption of recorded sound, reminding us that in 1926, synchronized dialogue for narrative films was hardly a foregone conclusion. Neither were our familiar understandings of sound recording and film sound at all obvious or unchallenged. Chapter 4 turns to the realm of sound "theory" as it was understood by engineers, technicians, and, later, by academics. Comparing the two sets of theorists, I develop a general set of questions and concerns as well as a set of practical problems that defined the assumptions governing sound research and technique, as well as the parameters of "good" sound practice.

Chapter 5 returns to many of the same issues, but this time focuses on the relationship between sound aesthetics, film form, technological change, and labor relations within Hollywood. Reexamining the studios' much-noted "delay" in adopting sound, I argue that competing sound theories and, more broadly, competing notions of cinematic aesthetics and narration, became an ideological battleground where different professional groups negotiated their workplace conflicts and resolved their competing aesthetics. Chapter 6 takes off from this same delay, but focuses on what was ultimately the central issue for the Hollywood sound film production—the question of sound space. Here I elaborate competing models of how one might construct acoustic space, and how each model embodies different understandings of cinematic representation, narration, and synchronization. As filmmakers gradually solved their conflicts and agreed upon a new and flexible set of formal strategies, they established the basic norms of sound and image that persist in large part today, shaping our own technologically mediated experience of the world. In the course of these arguments, I also hope to demonstrate that, like the world of visual representation, the world of sonic constructions has a history and a set of definable representational norms that deserve the same rigorous specification and interrogation that we are accustomed to apply to the history and theory of the image.

Much of the very best historical and theoretical work on film has been done all but oblivious to the relationship between technology and representation. Few contemporary critics even seem interested in questions of technology, unless they are addressed toward computers, digital imaging, or virtual reality. Unlike the words *culture, modernity, visuality*, or even *narrative*, the terms *technology* and *representation* seem to offer no immediate promise of compelling theoretical or historical argument, no promise of interdisciplinary relevance or groundbreaking conceptual development.

However, if we examine the cinema not only as a self-contained history but also as part of larger patterns of historical transformation, as we are accustomed to do with the printing press, for example, it is easier to see the potential gains offered by this approach. From this altered perspective, different questions appear not only as suddenly obvious but suddenly pressing, and familiar historical landscapes suddenly reveal alternate topographies. Unfortunately, this approach may also lead us to dissolve the cinema's specificity in a homogeneous solution innocuously dubbed "cultural history." I want to avoid this dissolution while still giving cinema's nonspecific characteristics their due; I hope, in fact, to use this broader view to give both the cinema and American modernity a *new* specificity.

Given what I take to be the absolute centrality of representational technologies to the emergence of the modern (and postmodern) world, I would argue that we are nowhere better able to track the relations between capital, science, and cultural practice than when we turn our attention toward photography, phonography, and the cinema. In order to do so, the historical analysis of representational technologies must avoid the various pitfalls so often attributed to it. Studies of technology have often been criticized for technological determinism, economism, and scientism, in their historical explanations, as well as for a more general methodological narrowness. In order to avoid these problems, a few tendencies of such studies need to be identified and actively resisted.

In a way, I wish to readdress the questions raised by the so-called "apparatus theory" of the 1970s and 1980s, and associated with such writers as Jean-Louis Baudry and Jean-Louis Comolli. Although each in his way attempts to focus upon the conjuncture between devices and particular cultural contexts, each winds up reinstating a kind of technological determinism. The perspective of the camera lens (and any type of lens) contains the entire humanist ideology of the Renaissance "inherent" in it.[6] Similarly, by rendering legible and spatially coherent images the camera produces the viewing subject as "transcendental" in a thoroughly Cartesian and Husserlian manner. Even if this claim were true (or provable), it is of such a general nature as to be true of nearly all images, perspectival or not. It certainly is not specific to the cinema. And while the camera (and the projection system that re-creates the movement of "objective reality")[7] is understood as the product of a particular epoch, that epoch is so broad (roughly five hundred years and counting) that whatever historical specificity it is supposed to convey is overwhelmed by the sameness of effect implied by each and every use of the cinema. Moreover, when Baudry pushes his analysis further and argues that the Unconscious is the originary apparatus producing *both* Plato's cave and the cinema, he so universalizes the cinema's effects as to render them utterly ahistorical.

Jean-Louis Comolli's approach, while more straightforwardly historical, nevertheless produces its own form of determinism. In an attempt to remain sensitive to the multiplicity and range of determinations on cinema technology, Comolli turns our attention away from "technicist" histories of the cinema that present endless lists of "firsts" and tell linear, chronological tales of successive "refinements." Instead, quoting Julia Kristeva, he calls for a stratified history "characterized by disjunctive temporality, which is recursive, dialectical and not reducible to a single meaning, but rather is made up of types of *signifying practices* whose plural series has neither origin nor end."[8] Unfortunately, Comolli's ideological determinations (while hardly the naïve "technicist" sort) essentially come down to one pressure—a demand for the "impression of reality."

Although reasonably filiating the cinema's brand of realism to prior forms in painting and theater, the demand remains diffuse, underdefined, and totalizing. Curiously technicist, however, when critiquing "empiricist" accounts of the history of depth of field, which ignore the ideological determinations upon its use and diffusion, he locates the entire effect of depth in the *lens* and ignores the equally relevant techniques of staging, lighting, aperture setting, and shutter speed. Even less convincing are his attempts to account for changes in apparent depth in, for example, the 1930s. At that moment, a general change from orthochromatic to panchromatic film stock is explained as more ideological than technological. "The hard and contrasted image of the image no longer satisfied the codes of photographic realism which had been transformed and refined with the growth of photography."[9] The addition of sound, he suggests, helped as well. Comolli offers no account of what, exactly, had changed about photography, nor why such a shift might have occurred. In the absence of any new research, Comolli's account seems even less plausible than those he critiques do.

Like Baudry's metapsychology, Comolli's analysis of depth of field, too, makes a great deal of hay from the fact that photographic lenses are ground to produce "Renaissance perspective." Absent from this argument is any account of the nature of the Renaissance *picture*. One gets no sense that Leon Battista Alberti insisted that a picture depict a significant human action, nor that such an action would likely be staged in a public space. Out of the complexity of pictorial practice (and Comolli is right that it is ideological through and through), Comolli selects only that element that can be located in the device. Consequently, and strangely like the "technicists" he deplores, he reifies whatever ideological significance cinematic space might have and renders it a function of the device.

While trying to turn our attention back to the basic questions asked by Comolli, I hope to do so in a more historically and theoretically circumspect manner. First, we must rethink technology by avoiding the reifications I discuss as the phenomenon of the camera's "click." The apparent punctuality and singularity of the gesture of button-pushing, so familiar to us all, may lead us to think of "photography" as a phenomenon based upon a particular device—the camera—and a particular act—pointing and shooting. It should be clear from even a moment's reflection, though, that the various processes embodied in the camera entail mixtures of quite distinct developments in chemistry, optics, mechanics, printing, and so on. It should also be clear that the act of button-pushing distracts us from the equally important acts of, for example, staging (even if we apparently choose not to stage), narrativizing, developing, printing, and presenting to the beholder. Although we leave many of these steps to others (to camera manufacturers or photo labs), they are crucial sites for representational intervention whose distinct character shapes an image's referential status, to name just one feature, in profoundly different ways.

The camera's click finds a methodological echo in the tendency to think of devices as immutable objects with given, self-defining properties. The discursive and even material parameters defining particular technologies, however, are always open to negotiation and redefinition. But standard histories still tend to view technology in terms of devices whose apparent material intransigence encourages us to regard their roles as stable and unchanging. Individual studies of specific media tell us just as repeatedly that their technological and cultural forms were by no means historical inevitabilities, but rather the result of complex interactions between technical possibilities, economic incentives, representational norms, and cultural demands. American history is littered with examples of how apparently stable objects like the telephone or radio, for example, underwent extraordinary redefinition in the course of their development in ways that were far more discursive and cultural than they were technical.[10] Moreover, these redefinitions often resulted in large-scale ramifications of the most profound economic and political sort, sometimes even decisively altering the possible forms of publicity available to the population.

With these concerns in mind, I suggest an approach that shaped both the course of my research and the structure of my arguments and examples. The terms *device*, *discourse*, *practice*, and *institution*, which serve as the defining parameters of my project, encompass material objects, their public reception and definition, the system of practices in which they are embedded, and the social and economic structures defining their use. And while I tend to privilege the discursive positioning and definition of devices over their physical forms, I balance that potentially idealist strategy with an equal emphasis on sedimented habits of representational practice, and the material constraints of technological capacity. Likewise, while it is tempting to see institutional settings, like the laboratory or the Hollywood studio, as ultimately determinate because they effectively impose specific limits on representational practice (and therefore representational form), institution is only one term in what might clumsily be called a four-term dialectic. Whichever terms seem most salient in a given chapter, the primary methodological precept is to look for mediation at each level. By operating in such a fashion, the many determinations that go into shaping, if never fully defining, a particular technological form become easier to identify, more specifically analyzable, and more clearly subject to critique and intervention.

As we move, however unwillingly, into the era of transnational and transcultural information flows, we are continually reminded of the absolute necessity of coming to grips with the role of technology in determining political, economic, social, and cultural life in both global and local scenes. One of our chief sites of struggle these days is, or ought to be, those arenas where the social definitions and implementations of technologies are determined, and a significant portion of our energy devoted to intervening, where possible, in those debates. Not every form of critique or intervention is valuable or progressive, however, and almost daily we read pointless condemnations of various technologies on the grounds that they are "ideological"—as if demonstrating a cultural or historical determination upon technological form were enough irremediably to taint it. Such criticisms, which find their roots in film theory's "apparatus criticism," remain largely unaware of the historical, philosophical, and cultural genealogies of current problems, and specifically ignorant of the manner in which ideologies mediately bear their fruit in concrete instances. More to the point, however, and all-too-frequently forgotten, is that such critiques threaten to discard the baby with the bath water.

The current rush to define the nature of digital technologies is a case in point. While the advent of digital images and sounds decisively shifted the terrain of debates about the relationship between the senses, technology, and representation, and give new meaning to my key terms *simulation* and *inscription*, new devices and new media are only part of the picture. When, today, everything can be digitally simulated (including those bits of "noise" or accident that Benjamin and Kracauer understood as the stigmata of history), it is tempting to believe that all bets are off, and that no image can refer in a reliable way to what it purports to depict. However, everyday epistemology tells us that reference depends as much, if not more, on politically and ethically defined *practices* of representation, and on habits of inference and abduction.

Nearly every contemporary newspaper image is digitized, and many are manipulated for purposes of clarity or rhetoric. We have not, however, consequently given up on the idea of photographic reliability since, in spite of our doubts about the medium, we still believe in the *institution* of journalism and the *practice* of journalists, which guarantee our faith far more than the simulacra produced by the computer. So, while it is all well and good to argue that every photograph is ideological to its core because every aspect of its form—framing, focus, perspective, range of tonalities—finds some precedent in the art of the Renaissance, and therefore in bourgeois humanism, or that every image has an "interpretive" or figural dimension, and is therefore never transparently documentary, such arguments go only halfway. Indeed, they lead to a kind of cynicism. No one today can be unaware that photos can be faked, yet that fact alone does not rule out the possibility that under *some* conditions a photograph might, indeed, tell the truth. And that is a possibility worth defending.

When lawyers in war crimes or police brutality trials turn the insights of ideological critique against our own political, cultural, and historical commitments, showing that technologically mediated reference is as unstable in the court as it is in the Hollywood movies we so love to deconstruct, we still too often flee, along with that mythical crowd at the Lumière brothers' first screening, from a representational actuality we too little understand. But we cannot. There is too much at stake, and it is too late to turn back from the media or from the powerful implications of such a deconstruction. We live in a world where experience is unavoidably mediated by photography, phonography, and motion pictures, and yet we still do not have an adequate account of how, in such a situation, a form may tell the truth, despite living every day of our lives in the faith that they do. Beginning the task of explaining this simple fact in all its contingency and complexity is the goal of what follows.

CHAPTER I

INSCRIPTIONS AND SIMULATIONS
The Imagination of Technology

I am experimenting upon an instrument which does for the Eye what the Phonograph does for the Ear. —Thomas Edison (October 1888)[1]

Was it really so clear only ten years after its first exhibition and mere weeks after its initial commercial availability what, exactly, the "Phonograph [had done] for the Ear?" And, assuming it was, in what sense could these acoustic transformations be duplicated for the eye? What made it appear obvious to discuss the two instruments as if somehow equivalent in effect? Given the numerous incommensurabilities between the visible and the audible, it seems strange to think of motion pictures and the phonograph in the same terms, and as accomplishing the same effects. Whatever their conceptual weaknesses, however, such cross-media and cross-sensory analogies were and continue to be intuitively appealing. We feel we know what Edison meant, even if it is difficult to describe precisely how his comparison illuminates the manner in which these instruments altered the shape of modern life.

Yet Edison was hardly the first to see analogies between different technological media. Before him, French photographer Nadar had imagined the possibility of an "acoustic daguerreotype," a sort of "box within which melodies are fixed and retained, just as the darkroom seizes and fixes images."[2] Twenty-two years before its material invention, he imagined that the phonograph would "do for the ear" what the photograph apparently had done for the eye. Indeed, there was hardly a writer in the nineteenth century, regardless of occupation, who did not speculate in similar terms. These revolutionary instruments had so touched the minds and imaginations of artists, scientists, and general public alike that no technological possibility seemed too remote, nor any new device too insignificant, to merit comment.[3] Even when based on misapprehension or outright fantasy, such vernacular discussion and evaluation conditioned not only contemporary understandings but also future technological innovation and exploitation, shaping the events through which new and old devices were reimagined, redefined, and even reinvented.

Both Edison's and Nadar's analogies try to describe the implications of one representational technology in terms of another's impact or, rather, as part of a broader project of historical transformation in which they jointly participate. Needless to say, the instruments imagined by both writers were yet to be invented, but like innumerable other forecasters, they felt intuitively that these devices were destined to change the nature, the context, the range, and the very possibilities of sensory experience. And in a very real sense, they did. Clearly, whatever was being "done for" the senses, or for experience in general, was a source of excitement and concern. Public conjecture, debate, and discussion were the primary outlets through which the culture at large, as well as specific subcultures, negotiated and defined the impact and meaning of these new representational forms.

Edison himself felt compelled to address the mass interest generated by the phonograph in an 1878 essay, where he tried to define the impact of a device that had "commanded such profound and earnest attention throughout the civilized world." This attention, he muses, is due to "that peculiarity of the invention which brings its possibilities within range of the speculative imaginations of all thinking people, as well as the almost universal applicability of the foundation principle, namely, the gathering up and retaining of sounds hitherto fugitive, and their reproduction at will."[4] He adds further that,

> From the very abundance of conjectural and prophetic opinions which have been disseminated from the press, the public is liable to become confused, and less accurately informed as to the immediate result and effects of the phonograph than if the invention had been one confined to certain specific applications, and therefore of less interest to the masses.

Edison's mixture of satisfaction (mass interest means mass market) and trepidation (mass conjecture means the phonograph's future is beyond his immediate control) indicates both the limits and the potential of public debate. Such bursts of anticipatory discussion were typical of a technological utopianism permeating nineteenth-century American culture and characterized the reception of most of the century's representational technologies—even currently unavailable ones.[5] While the material invention of instruments was still in the hands of a few, their imaginative invention and reinvention were profitably and dangerously in the hands of the many, who might expand their significance and their uses beyond an inventor's wildest dreams or even against his wishes.

Indeed, Edison's own invention was anticipated by several months by a friend of Verlaine, Manet, and Baudelaire—the poet Charles Cros. This dilettante inventor deposited a sealed envelope detailing his phonographic device with the Académie des Sciences on April 30, 1877, some eight months before Edison received his own patent.[6] Described in nearly every history of the phonograph as "a dreamer," Cros's success at *conceiving* the phonograph—there is no doubt it would have worked—demonstrates just how involving these questions were to the imagination of the day, and how those concerned with giving poetic utterance to the newest forms of modern life were equally engaged with its more technical expressions. But it is Cros's close friend, Villiers de l'Isle-Adam, who is better remembered for his less scientific but more compelling and disturbing musings on the meaning of Edison's work.

Begun in 1878 and published in 1889 (one of the first copies being given to the inventor while at the Exposition Universelle), Villiers's *L'Eve future* offers us a Faustian Edison, who brings his immense technical knowledge to bear on the most difficult of mechanical problems. Entertaining a visiting old friend, Lord Celian Ewald, the "Phonograph's Papa"[7] listens attentively to his companion's tale of woe. Ewald laments that he has fallen in love with a singer, Alicia Clary, the most beautiful woman in the world, whose one fault, alas, is an utterly vulgar soul. Ewald regards her physical beauty as quite literally ideal and her soul just as surely profane, dominated as it is by stupidity and petit bourgeois materialism. Edison is spurred by this account to complete a languishing project—a mechanical woman—which he offers as a substitute for Alicia. The skeptical Ewald unwittingly falls in love with Hadaly, an automaton constructed in Alicia's physical image, but who expresses sentiments and ideas he finds profound and irresistible. Mechanical to the core, the ghost in this machine is a pair of phonographs cleverly designed to allow her to respond intelligently and with apparent feeling to any number of questions or occasions. Ecstatic over what he believes to be an ideal combination of body and soul, Ewald sets sail for home with Hadaly, only to lose his love to a violent storm at sea.[8]

Beyond the equally stunning literary experimentation and virulent misogyny, *L'Eve future* offers a striking portrait of an Edison ("The Wizard of Menlo

Park," not the historical individual, Villiers tells us) whose various inventions (the author attributes the telephone, the microphone, the phonograph, and the light bulb to him) haunt the world with alluring, fantastic, and dangerous possibilities. Like his friend Cros, Villiers offers an unrealized vision of technological possibility as well as technological consequence, imagining the next logical step on the transformative road of modernity—a vision of creeping calculation and mechanism, affecting not only industry but the human soul as well.[9]

Edison himself offered a different view of the possibilities and consequences implied by his device in an essay for the *North American Review*, in June 1878, when, by some accounts, Villiers was already hard at work upon an early version of *L'Eve future*. The picture Edison presents is less startling, but no less troubling. It is worth quoting at length. "The . . . stage of development reached by . . . the phonograph," he notes, "demonstrates the following *faits accomplis*":

1. The captivity of all manner of sound waves heretofore designated as "fugitive," and their permanent retention.
2. Their reproduction with all their original characteristics at will, without the presence or consent of the original source, and after the lapse of any period of time.
3. The transmission of such captive sounds through ordinary channels of commercial intercourse and trade in material form, for purposes of communication or as merchantable goods.
4. Indefinite multiplication and preservation of such sounds, without regard for the existence or non-existence of the original source.
5. The captivation of sounds, with or without the knowledge or consent of the source of their origin.[10]

It takes but a moment's reflection to recognize these *faits* as classic philosophical criteria, deriving from Plato's *Phaedrus*, for differentiating writing from speech—its relative "absence" compared to the presence of speech, its reuse in new contexts "without the knowledge or consent of the" speaker, or its "indefinite multiplication . . . and preservation" regardless of the existence or nonexistence of the speaker. More capitalist than Platonist, however, Edison finds little to criticize here, questions of profit far outweighing questions of Being. Phrases like "without the presence or consent of the source," "after the lapse of any time," and "preservation . . . without regard for the existence or non-existence of the original source" have a utopian dimension to be sure, promising future generations the possibility of hearing long-passed statesmen, singers, and poets in their own voices, but they sound an ominous note as well. The words "without knowledge or consent" in particular evoke a situation for which the word "captivation" is all-too-perfect.

Revisiting these concerns ten years later in the same magazine, Edison describes the "Perfected Phonograph" as having fulfilled most of the promises he made in his earlier essay, but he nevertheless alters a few of his claims. Among the notable changes in his list of phonographic possibilities is an elaboration of point five, above. Now eschewing a language redolent of surveillance and punishment, he instead emphasizes its capabilities as an unimpeachable, *legal* record of business dealings. Surveillance, while still implicit here, emerges explicitly at the end of his essay, where he suggests that we will or should internalize a psychic phonograph—an external ear—whose possibly real ubiquity "will teach us to be careful what we say—for it imparts to us the gift of hearing ourselves as others hear us—exerting thus a decided moral influence."[11]

Still, the analogy between recording and writing is neither univocal nor unfailingly pessimistic, and Edison's vision of a perpetual and anonymous surveillance is by no means the only interpretation available. But just as with Villiers's automaton, who threatens to confuse previously secure definitions of humanity, recorded sound, as "writing," threatens to disrupt long-established patterns of human interaction and to create new forms of experience for which there are no clear historical precedents. From the nineteenth century forward, representational technologies increasingly shaped and even dominated certain arenas of modern life, producing what Walter Benjamin would later call new "structures of experience," whose consequences for individual and collective life could not easily be foretold. So important did these consequences seem, however, that an entire culture groped collectively for an appropriate vocabulary through which to express them.

Therefore, when Edison turns to a logic of writing and Villiers to one based upon the imitation of human functioning in order to comprehend the phonographic transformation of life, these tropes, metaphors, or ways of thinking are not to be taken lightly. They serve to mark the manner in which individuals, here ones endowed with a fair amount of cultural prestige, struggled to understand and control the world in which they lived. But the importance of these tropes—inscription and simulation—cannot be measured by their authors' prestige alone. Rather, they need to be evaluated in terms of their cultural pervasiveness, their explanatory power, and the degree of their historical influence and institutionalization. In short, did they make material differences in nineteenth-century lives, and does their metaphorical power translate into historical explanation?

In this case they certainly do, for writing and physiological or perceptual simulation are, in fact, the master tropes through which the nineteenth century sought to come to grips with the newness of technologically mediated sensory experience, and together they set the boundaries for describing, understanding, and deploying new representational technologies. Because of their flexibility, their ability to concretize a wide variety of real dilemmas in communicable form, and because they organized both logics of representational production and interpretation, writing and simulation were, and continue to be, powerfully influential conceptual tools whose assumptions are readily translated into practical strategies.

Why do notions of writing and of human simulation so pervade discussions of technological representation across media? On the one hand, the answer is quite simple—even obvious. The link between representation and human perception is perhaps the single most persistent mechanism for explaining representational form in all of history. To say that a painting duplicates what the painter sees reaches back to our earliest discussions of art, and straddles otherwise distinct traditions. Nevertheless, no painting ever *does* duplicate vision, nor is perceptual duplication the only way to discuss image-making.[12] The idea of representation as a form of perceptual simulation, while a continually available metaphor, comes to the fore at specific historical moments in order to justify particular representational practices or to solve particular cultural problems. If nothing else, the analogy describes a very clear and concrete manner for linking representation to human experience.

If, for instance, one understands the telephone as a "perfection" or "extension" of the ear, rather than as a radically new mode of communication, it seems not only more familiar but more easily lends itself to particular social uses. Understood as a substitute for the face-to-face relationship of interlocutors, the telephone is more likely to be integrated into daily life as a means for personal communication. Understood otherwise, as it was in Hungary until the 1930s, the telephone becomes a medium for public communication on the model of the radio, where listeners, following a schedule, would simultaneously pick up their phones to hear concerts or the news.[13] By the same token, understanding a painting as a record of the artist's sight implies different norms and criteria of evaluation than if one understands it as inherently symbolic or rhetorical.

By contrast, metaphors of writing imply the concepts of inscription and storage, as well as a medium's relationship to language and legibility. Taken as a form of writing, phonography's ability to store speech and transmit it across space and time seems less disruptive, and more easily masterable. Likewise, the analogy suggests particular uses over others. Indeed, Edison clearly assumed (and relentlessly promoted the idea) that phonograms might best be employed as substitutes for business letters and as filing and retrieval systems. Oddly enough, he initially seems almost oblivious to the possibility that the phonograph might record music, and thereby serve as a kind of prosthetic ear. Yet tropes of inscription do not *preclude* perceptual models; they sometimes come to their aid, helping mediate the relationship of perception to technological forms of sensory experience. Together, they help sort out the meaning of sensory apparatuses that promised inhuman sensory abilities, such as ears or eyes that could *store* experience, offering the possibility of a sense with a memory.

In the case of the nineteenth century, particularly in the United States where these technologies found almost immediate employment and massive dissemination, the conceptual pair inscription/simulation worked above all to normalize a new series of conditions that had come to define the technologically mediated relationships between perception, representation, and meaning. As Villiers suggests, the phonograph stood emblematically for the fact that speaking, singing, and music-making no longer required the presence of a human performer. Exacerbating trends begun with the telephone, the phonograph had made it quite literally possible to hear the voices of those not present, and what is more, no longer alive. The device outstripped human abilities in other ways, too. As Edison never ceased to insist, it was suddenly possible to speak with people on the other side of the globe or those far in the future. Unlike human ears and voices, the phonograph could *record precisely* what had been said, preserving not only words but the most contingent nuances of their expression, and then reproduce them at will, again and again, with no variation.

But more than simply altering the parameters of acoustic experience, the phonograph, like the earlier photograph and the later cinema, promised to alter the human relationship to phenomenal experience and to history more generally, preserving in a permanent form those events "heretofore designated as 'fugitive.' " Not only could these records be stored like manuscripts in an archive, they could and did become commodities, so that formerly singular acoustic experiences were available for mass consumption and private possession. Perhaps most alarmingly, human intervention in the process of representation became, in a manner of speaking, literally unnecessary. It was frequently likened to photography, which had long been described as images "drawn by sunlight itself," as a process that "gives [Nature] the power to reproduce her-self," as "nature copying nature, by nature's hand," or as "The Process by Which Natural Objects May be Made to Delineate Themselves without the Aid of the Artist's Pencil."[14] With the camera and photochemistry, images appeared to have become "autographic"—capable of inscribing themselves (figure 1.1). Indeed, if human intervention played a role, it was to be voluntarily "passive." So thoroughly did this idea permeate the popular imagination that it became a commonplace of melodramatic narratives like the plays *The Octoroon* and *The Phonograph Witness*, the novel *Captain Kodak*, and the short story "A Strange Witness," for a camera, a phonograph, or a motion picture camera "acciden-tally" to record a murder and thereby indict the guilty without prejudicial bias in the dime novel's denouement or the play's final act.[15] Such beliefs conse-quently fueled other concerns, since, if hearing and seeing could be divorced from human bodies, the possibilities for surveillance expanded dramatically.[16]

Still more troubling, these devices decisively altered the boundaries between the visible and the invisible, the audible and the inaudible. Time and again, representational technologies are described as "more perfect" than human senses, able to "make up for" previously unnoticed "deficiencies."

When Étienne-Jules Marey developed his "graphical method" for the sci-ences, he repeatedly described his cameras and related devices as instruments

FIGURE 1.1
Typical form of "autographic" inscription—the movements of the vibrating fork pro-duce their own record.

designed to amend and even replace our inferior senses, which were too weak to perceive, and therefore accurately study, nature in motion.[17] Optical and acoustical technologies were therefore commonly envisioned as prostheses, i.e. simulations of a sort, extending the range and accuracy of sensory organs too coarse or too limited to meet the demands of modern life.[18] Simultaneously liberating and threatening, these prosthetic senses made available an entirely new world of experience, but elicited compensatory fantasies of either sensory inadequacy or sensory omnipotence. It is but a short imaginative leap from the prosthetic eyes and ears of science to the monstrous automata of Villiers and E. T. A. Hoffmann, and all they imply.

While metaphors of writing and of human simulation did not arise simply in response to technological transformations, they flourished because of them. Such habitual ways of thinking about technologies had circulated for some time and arose not through philosophical speculation about new forms of experience, but out of the concrete work of representation in and around the laboratory. The nineteenth-century pairing of graphical and simulationist tropes has a long and deeply sedimented history of its own, a history that explains not only why these forms of expression dominate explanations of new technologies, but how they determined the very form of representational instruments themselves.

One of the earliest and most illuminating discussions of the phonograph, by American physicist Alfred M. Mayer, exemplifies the salience of these two ideas within a quasi-scientific context. Mayer's title, "On Edison's Talking-Machine," reminds us that, as contemporary essays and even periodical indexes confirm, the category "talking machine" covered a wide and intriguing semantic field. Mayer writes, "All talking-machines may be reduced to two types. That of Prof. Faber, of Vienna, is the most perfect example of one type; that of Mr. Edison is the only example of the other."

> Faber worked at the source of articulate sounds, and built up an artificial organ of speech, whose parts, as nearly as possible, perform the same functions as corresponding organs in our vocal apparatus. A vibrating ivory reed ... forms its vocal chords. ... There is an oral cavity. ... A rubber tongue and lips make the consonants ... and a tube is attached to its nose when it speaks French. This is the anatomy of this really wonderful piece of mechanism.[19]

Faber, then a well-known performer (the title "professor" probably alluded to his mechanical skill and mastery over the intricate device), had constructed what was universally regarded as the most successful of the previous century's many talking automata. The Euphonia, as it was known, had startled and impressed audiences since 1846 by imitating human speech and answering impromptu questions in several languages under the control of Faber's keyboard and bellows.[20] Unlike previous talking machines, it articulated clearly and required no ventriloquist or hidden speaking tubes—the machine could actually be made to talk intelligibly.[21] (See figure 1.2.) Despite being praised for his ingenuity, Faber died in poverty, audiences apparently preferring contraptions like the infamous "invisible girl" that were based on some form of deception. (See figure 1.3.)

In 1873 Faber's device joined Barnum on tour in the United States and, depending on one's information, was either disappointingly unsuccessful or reasonably profitable.[22] Regardless of its monetary success or failure, however, Faber's machine was, by the time of Mayer's essay, well known in

FIGURE 1.2
Faber's Euphonia—"Breath" is supplied by the bellows. Mouth, tongue, and larynx are controlled by the keyboard.

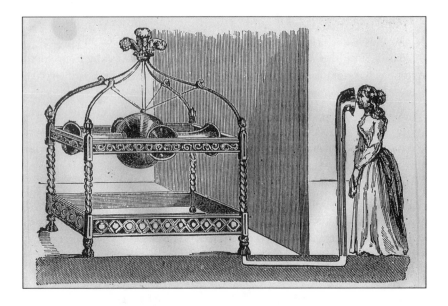

FIGURE 1.3
The "Invisible Girl."

America and even considered of comparable importance with Edison's. The comparison itself, however, remains somewhat surprising. The persistent pairing of quasi-scientific instruments with what, from our perspective, can only seem fairground attractions, is absolutely fundamental to the popular understanding of talking machines in the 1800s. Indeed, the very designation "talking machine," rather than "phonograph" or "music machine," is instructive,[23] and this unlikely categorization was even central to the *scientific* understanding of sound representations, where the boundary between ingenious but vaguely disreputable entertainment devices and "genuine" scientific instruments was not only permeable but almost necessarily so.[24] (See figure 1.4.)

Mayer's decision to divide "talking-machines" according to the Faber/Edison split is entirely consistent with the two basic types of sound representation typical of nineteenth-century acoustics. Describing the difference between their approaches to the problem of vocal representation, Mayer argues,

> Faber attacked the problem on its physiological side. Quite differently works Mr. Edison: he attacks the problem, not at the source of origin of the vibrations which make articulate speech, but, considering these vibra-

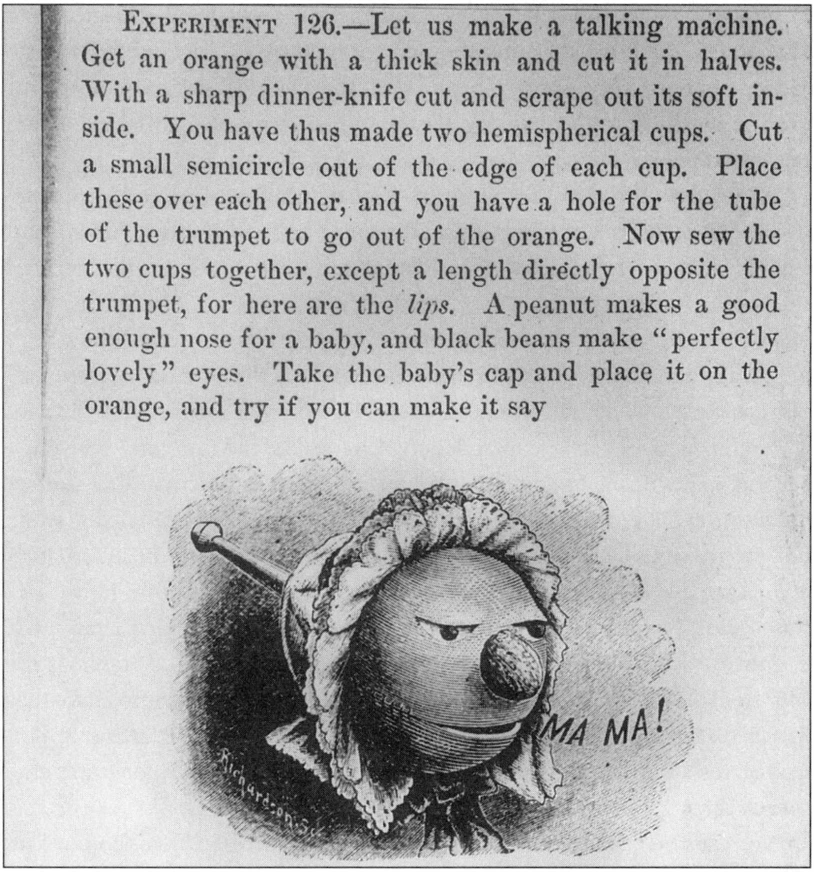

EXPERIMENT 126.—Let us make a talking machine. Get an orange with a thick skin and cut it in halves. With a sharp dinner-knife cut and scrape out its soft inside. You have thus made two hemispherical cups. Cut a small semicircle out of the edge of each cup. Place these over each other, and you have a hole for the tube of the trumpet to go out of the orange. Now sew the two cups together, except a length directly opposite the trumpet, for here are the *lips*. A peanut makes a good enough nose for a baby, and black beans make "perfectly lovely" eyes. Take the baby's cap and place it on the orange, and try if you can make it say

FIGURE I.4
Physicist Alfred Marshall Mayer suggested that schoolchildren construct and *decorate* a "talking head" as the final project in a series of experiments designed to teach the principles of acoustics.

tions as already made, it matters not how, he makes these vibrations impress themselves on a sheet of metallic foil, and then reproduces from these impressions the sonorous vibrations which made them.

Faber solved the problem by reproducing the mechanical *causes* of the vibrations making voice and speech; Edison solved it by obtaining the mechanical *effects* of these vibrations. Faber reproduced the movements of our vocal organs; Edison reproduced the motions which the drum-skin of the ear has when this organ is acted on by the vibrations *caused* by the movements of the vocal organs.[25]

In treating sound purely as vibrations to be transcribed, Edison effectively severed recorded sound from any human or even necessarily external origin, raising the possibility (soon to be realized) that phonographic sounds might even be produced *without an original or prior sound*, simply by making indentations on the foil cylinder.

While it would be disingenuous to claim that Mayer's dichotomy precisely duplicates the writing/simulation division I extrapolate from Edison and Villiers, they overlap in several essential ways. By ignoring questions of source and focusing on what Mayer calls "effects," Edison's device effectively embodies several of the basic features of writing. First, it does away, both practically and, more important, conceptually, with the necessary "presence" of a speaker. The "actual" source of a sound is irrelevant to its representation, and significantly the necessary link between speaker and utterance is severed. Second, like writing, and unlike Faber's automaton, it functions as a storage medium, archive, and means of exact repetition or citation. Finally, while both approaches to sound representation raise the specter of the inhuman in an arena, speaking and hearing, which previously could only have been thought of as essentially human, Edison's device does so in the manner that linguistic writing does, not in the manner of Faber's "Frankenstein Monster."[26] Perhaps most important for our purposes is the simple fact that Mayer distinguishes the phonograph not from other devices of inscription like Marey's but from a speaking automaton—a mechanism for producing speech through physiological simulation.

Nor was the writing/simulation dichotomy exclusive to discussions of the phonograph. The inventor of the century's other major talking machine, Alexander Graham Bell, brings together the importance of writing and simulation in a striking manner, and indirectly suggests their historical interconnections. Looking back on his own career in 1922, Bell tries to account for his intellectual formation, searching for the key events and conditions that might have led him to develop the telephone. Two strands of influence are of particular importance to him, and he expounds upon them at length. First, he describes the important influence of his father, a speech teacher who

> devised a remarkable system of symbols for depicting the actions of the vocal organs in uttering sounds. . . . He claimed that what he had really invented was a universal alphabet, capable of expressing the sounds of all languages in a single alphabet, and that his letters, instead of being arbitrary characters, were symbolic representations of the organs of speech and of the way in which they were put together in uttering sound.[27]

The idea of a nonarbitrary writing profoundly structures the understanding of phonography throughout the century. In fact, the very term "phonography" initially referred to a stenographic system developed by Isaac Pitman in 1837, which, by transcribing *sounds* instead of words, was expected to offer a more direct, almost analogical form of writing.[28] Yet another stenographer (and typesetter), Edouard-Léon Scott de Martinville, searching for his own method of automatic writing, developed the direct mechanical precursor to the Edison phonograph—the *phonautographe*—a device he described as allowing sound to "write itself" automatically.[29] (See figure 1.5.) While indeed capable of inscribing vibrations on a cylinder covered in lampblack, the device could not be used to reproduce sound (although it was nearly identical to Edison's instrument), and Scott, instead, expected it to foster a new mode of reading, which would decipher the instrument's squiggles.[30] (See figure 1.6.)

Even more illuminating are the repeated comparisons made between the new media, especially phonography, and hieroglyphic writing. Like the phonograph, it was argued, the hieroglyph's nonarbitrary or iconic aspects rendered it a more likely candidate for the status of "universal language." In both Bell's work and Edison's this is even an explicit goal. Significantly, a hiero-

FIGURE 1.5
The *Phonautograph*.

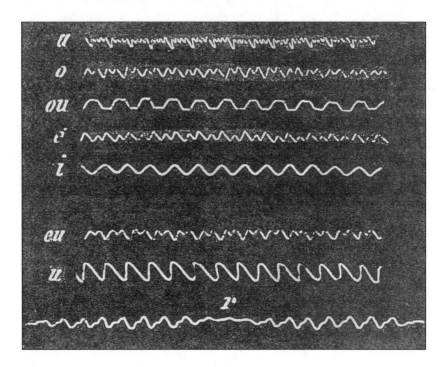

FIGURE 1.6
Phonautographic inscriptions of vowel sounds.

glyphic model of writing overcomes some of the "problems" Edison signals in his essays by creating a causal, or as C. S. Peirce might say, existential link between sign and object. And as Thomas Levin notes vis-à-vis Theodor W. Adorno, the "autographic" character of phonographic signs reawakened the connection between music and writing, just as for Marey the "graphic images [produced by the body were] the phenomenon's own language."[31]

Despite enormous differences in material form and mode of transcription, stenographic and phonographic systems of recording owed their special status to their capacity to produce something like an analogical, continuous, and autographic script, capable of transcribing any spoken language. Describing his role as a performer in his father's demonstrations of the system, Bell adds an important variation on this motif. Alexander, it seems, would leave the room while an audience member suggested a sound to his father. Bell père would "write" it in his system of "Visible Language," and Alexander, returning to the room, would decipher it. Bell recalls that on occasion he would read the signs and, reproducing their physiological instructions exactly, produce unusual, nonlinguistic noises, which were always greeted by applause. As he notes, "It was just as easy for him to spell the sound of a cough, or a sneeze, or a click to a horse, as a sound that formed an element of human speech."[32] This was truly a "universal alphabet."

Edison's device, as described by Mayer, presumes a precisely analogous ability to record and reproduce sounds indiscriminately. Since for Edison the entire process was a question of recording vibrations qua vibrations, it mattered little what the *source* of the vibrations was—unlike talking heads, which could only reproduce speech. In fact, it was a commonplace of early reports of the phonograph (and the photograph and cinema, too) to give an arbitrary listing of the endless variety of sounds that might be reproduced by the device and to remark upon the novelty of this fact. Not only could the phonograph "speak" (as could Prof. Faber's Euphonia), it could duplicate brass bands, opera, "artistic whistling," "roosters crowing, ducks quarreling, turkeys gobbling," and even babies crying.[33] In effect, and with similar mechanical perfection, young Bell himself had become a phonograph—a reproducer of nonarbitrary inscriptions—acting as the playback mechanism for his father's recording apparatus.

The other major source of Bell's interest in talking machines is both more curious and more illuminating since it suggests a genealogy and an institutional context for tropes of bodily simulation. At some point during his early youth, Bell was given the opportunity to meet with Sir Charles Wheatstone, known at the time for his work with electricity and the telegraph, and for his research in stereoscopic photography. But as Bell remembers the meeting, the discussion had nothing to do with these marvels. Instead, it dealt with Wheatstone's interest in Baron Wolfgang von Kempelen:

You have probably all heard of the celebrated automaton chess player of the Baron von Kempelen, which appeared in the eighteenth century and startled all Europe by beating the most celebrated chess-players on the Continent. The story has come down to us that a dwarf was concealed in the apparatus, who guided the machinery and dictated the moves.

Many persons have imagined that the Baron's equally celebrated automaton speaking machine, which was said to have uttered words and sentences in a childish voice, also constituted an imposition on the public; but, on the other hand, there were some grounds for believing that this might have been a real automaton after all, for the Baron von Kempelen published a book upon "The Mechanism of Human Speech," in which he gave a full description of his speaking machine, with copious illustrations.[34]

Largely forgotten today,[35] Kempelen was one of the mechanical geniuses of his age, and one of its great illusionists. While Bell gets a few details wrong, Kempelen *was* the designer of an internationally renowned chess-playing "automaton" that did conceal an expert player within its workings, and he *was* widely praised both for his talking machine (something of a precursor to Faber's) and for his important work in phonology. (See figures 1.7, 1.8, and 1.9.) The most important aspect of this anecdote for us, however, lies in the impact the speaking machine had on Bell, and how the idea of human simu-

FIGURE 1.7
Kempelen's chess-player (note generic resemblance to Faber's Euphonia, fig. 1.2).

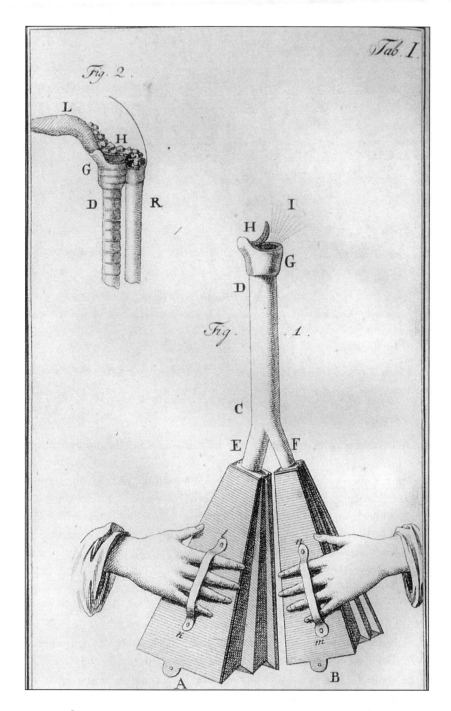

FIGURE 1.8
One of Kempelen's speaking machines, designed to simulate human lungs, trachea, larynx, and glottis. (Photo courtesy Department of Special Collections, the University of Chicago)

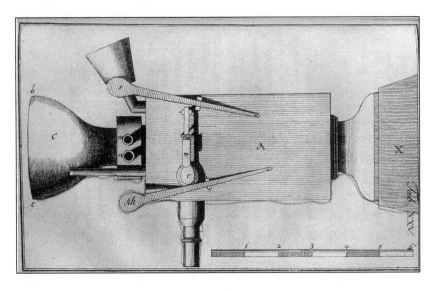

FIGURE I.9
Kempelen's successful speaking machine. Copied by Charles Wheatstone and Alexander Graham Bell. (Photo courtesy Department of Special Collections, the University of Chicago)

lation became a formative influence on his thinking about the representation of sound.

Encouraged by the limited success Wheatstone had in duplicating Kempelen's device, Bell borrowed the book, "devoured" it, and, with his father's encouragement, set to work with his brother on their own talking machine. "We divided up the work between us, his special part consisting of the larynx and vocal chords, to be operated by the wind chest of a parlor organ, while I undertook the mouth and tongue." This was no whimsical endeavor, he later remembered.

> My father took an extraordinary interest in the proposed talking-machine and encouraged us in every way. I now realize, as I could not then, that he looked upon the machine as a valuable educational toy, which would compel us to become familiar with the operation of the vocal organs, quite independently of any practical results attained. This accounts for the fact that he did not encourage us to follow in the footsteps of Kempelen and Wheatstone, but rather sought to have us copy Nature herself.[36]

In fact, their mimetic impulse took them as far as the slaughter of a neighbor's cat, which they perpetrated in a quest to secure a larynx for their machine. Although a bit extreme in their methods, in the technique of physiological simulation they were, as their father understood, following one of the scientific precepts of the age.

The simulation of bodily form and function went far beyond attempts to duplicate the organs of sound *production*, however. As is by now clear, the human sense organs—that is, the devices for *receiving* sounds—were also the subject of vigorous research and innovation. Scott's initial report to the Académie was spurred, for example, by the news of a foreign experimenter who, with the aid of an instrument designer, had designed a method "for the automatic inscription of the vibratory movements of the middle-ear apparatus of a freshly decapitated dog."[37] Others sought to produce vowel sounds by taking plaster casts from the larynxes, throats, and mouths of human cadavers and duplicating them in wood, leather, or rubber.[38] (See figures 1.10 and 1.11.) Perhaps most familiarly, innumerable scientists and inventors duplicated human binocularity in their design of stereoscopic cameras and viewers, and philosophers and scientists, tracing their lineage to Descartes, dissected eyes to build cameras and microscopes.

As the cases of Bell, Scott, and even Edison demonstrate, understanding sound representation as a form of writing and understanding it as a process of duplicating human bodily capacities were not mutually exclusive. Part of the reason they appear so prominently in the literature on technological representation is that they emerge as explanations and defenses against the various questions these instruments raised about the nature of sensory experience. Each motif serves in a different way to grasp, articulate, and sometimes defuse troubling aspects of the emerging and ever-expanding experiential revolution. Still, they are not simply interchangeable. While they often function as a pair, each belongs to a different historical and institutional trajectory. It is crucial, therefore, to demonstrate both how the pair emerge as discursive mechanisms with specific roles to play, and how those discourses and their assumptions filter through the culture at large, coming to bear on individuals and their dealings with new representational forms.

In short, I want to demonstrate not only that vernacular discourses about new representational technologies helped to bring them into the realm of common understanding, but that these same discourses helped to determine the very form of those devices. Additionally, and as a consequence, I will show how they came to shape basic ideas about representation in general, so

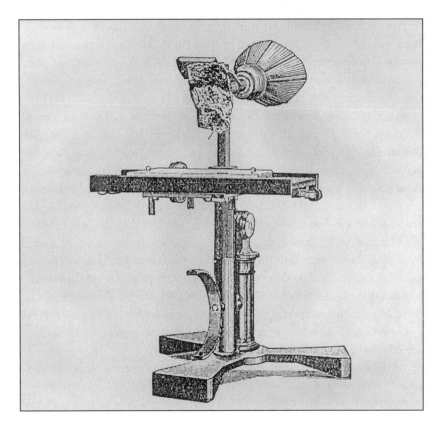

FIGURE 1.10
Bell's "Ear Phonautograph," 1874. Constructed from dissected human auditory organs.

that while I explore the history of the technological representation of sound, my claims apply to vision as well.

If the dichotomy between inscription and simulation is as important as I contend, serving not only to explain technological representations but also to generate and guide the development of new devices and new paradigms of representational practice, we need to understand its rootedness within the field of acoustics as it was taking shape in the late eighteenth and early nineteenth centuries. While the science of acoustics received its first impetus from the study of music and the theory of harmony, its other grand project and reigning obsession was the analysis and reproduction of human vowel sounds. From the outset, the attempt to reproduce and to explain

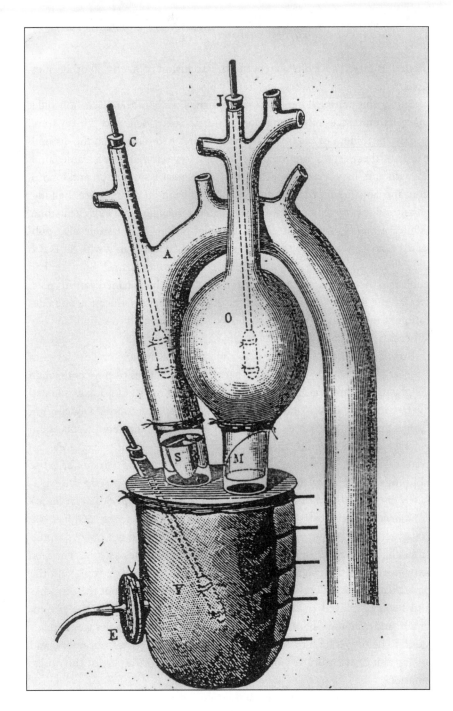

FIGURE I.II

Artificial Heart constructed by Étienne-Jules Marey in order to study coronary function via "mimetic experiments."

them dominated sound researches to the exclusion of nearly all other questions.[39]

Still, if this pairing emerges from the formal study of the voice, how did it make its influence felt among nonscientists? Scott, Bell, and Edison were, strictly speaking, not scientists, yet each figures prominently in the dissemination of these representational paradigms. How is it that they became familiar with and, in a sense, respected the paradigms of simulation and inscription, and came to internalize them as habits of thought and practice? Bell suggests the most plausible answer—through figures like Charles Wheatstone. While primarily a scientist and instrument-maker, Wheatstone also published essays for general audiences and famously collaborated with Sir David Brewster on a widely read book about vision and stereoscopy.[40] Brewster had published the popular *Letters on Natural Magic*, which explained various spectacular and illusory phenomena in the interests of demystification and education.[41] Several chapters of Brewster's book are devoted to sound, and like Wheatstone, he spends time discussing Kempelen's work.

Brewster's *Letters* is only the best known of a seemingly endless series of popular scientific texts that sought to teach the principles of physics through delightful and often astonishing demonstrations. Brewster's book was joined by (to name but a few) John Ayrton Paris's *Philosophy in Sport Made Science in Earnest* (1st ed., 1827; 8th ed., 1857), John H. Pepper's immensely successful *Playbook of Science* (more than five editions) and *Pneumatics and Acoustics: Science Simplified* (c. 1888), Gaston Tissandier's famous *Popular Scientific Recreations* (1883, 1890), and Alfred M. Mayer's own *Sound: A Series of Simple, Entertaining, and Inexpensive Experiments in the Phenomena of Sound for the use of Students of Every Age* (1891).[42] Typically, "recreational science" involved the construction of "philosophical toys," or devices designed not only to amuse and delight but to embody and teach scientific principles as well. The best known include the kaleidoscope, the thaumatrope, the zoetrope, and the phenakistoscope, all considered precursors to the cinema, but the term has also been employed to describe devices like the stereoscope, the phonograph, and even the cinema itself. In short, sensory technologies were understood to belong at least partially to the realm of instructive amusement, and an account of their underlying principles was therefore a part of their presentation.

Whatever the specific circumstances, a great many people learned about sound, about vibrations, and about talking machines through mediators such as the philosophical toy, and its human counterpart, the lecturing experimentalist.[43] Of equal importance later in the century were magazines such as

Popular Science Monthly, Scientific American, the *Atlantic Monthly,* and the *North American Review.* Uniting these venues was a shared emphasis on the pleasures of learning about advances in science and the development of intriguing, if only marginally scientific, devices. Wheatstone made many contributions to this mediating arena, including the invention of a few of his own philosophical toys. Typical of our scholarly tendency to privilege sight over sound, however, the Wheatstone stereoscope is well known, while his kaleidophone is largely forgotten.[44] Comprised of a few metal bars, whose differently oriented tips were covered in luminous paint or highly reflective metal, it could be vibrated with a violin bow and the resulting patterns seen as bright figures traced in the air. Despite its simplicity, it persisted well into the 1890s as a device for demonstrating principles of vibratory motion. Wheatstone's interest in acoustics, however, went well beyond this device and many of his earliest publications concern acoustical phenomena.[45]

Particularly illuminating in this regard is an article he contributed to the *London and Westminster Review* in 1837.[46] Entitled "On Reed Organ Pipes, Speaking Machines, etc.," it took the occasion of a review of three books on sound to explain basic principles of acoustics and to discuss the current state of research. The three works reviewed are Kempelen's *Mécanisme de la Parole,* C. G. Kratzenstein's *Tentamen Coronatum de Voce,* and Robert Willis's *On the Vowel Sounds and on Reed Organ Pipes.* More important than what Wheatstone says about any of them in particular is the simple act of juxtaposition. Kratzenstein and Kempelen had crossed paths before, when the former claimed the prize offered by the Imperial Academy of St. Petersburg in 1779, for the first researcher to successfully reproduce the five basic vowel sounds. Kratzenstein, a professor of physics and medicine at Copenhagen University, barely outpaced Kempelen, who had also begun an important investigation of the mechanism of speech—specifically, the production of vowels. The results of Kempelen's research were his book and his famous speaking machine.

Robert Willis was also destined to cross Kempelen's path, although after the latter's death. As Bell recalls, Kempelen's other claim to fame was his notoriously successful chess-playing automaton, famously debunked by Edgar Allan Poe in an essay, "Maelzel's Chess Player."[47] Poe, however, was by no means the first to expose the illusion, and one of his most important precursors was a twenty-year-old Englishman named Robert Willis. Willis would achieve renown later in life for his important work in physics, where he is principally remembered for having, effectively, brought the study of the voice firmly within the boundaries of acoustics.[48] He may also be familiar to students of nineteenth-century visual culture as one of the originators of what Étienne-Jules Marey was later to dub "the graphical method" of physiological inscription.[49]

In this single review, Wheatstone manages to make, however unwittingly, important connections between acoustics, graphical inscription, and automata, implicating them in a shared project of research into the mechanical representation of sound. The graphical and physiological approaches to the representation and analysis of sound would continue to compete and to clarify one another for years, culminating in the opposed but complementary researches of Hermann von Helmholtz and Rudolph Koenig. The former was a die-hard physiologist whose experiments and explanations were always based on concrete, human, anatomical details, and the latter, the century's most important maker of acoustic instruments,[50] pursued vibrations independent of their supposed source, inscribing their traces in a variety of media. While, together, they ultimately helped prove the "fixed-pitch" theory of vowel formation, each assumed a different guiding notion of representation and, therefore, of evidence and proof. Koenig cast his lot with Marey and company, Helmholtz, oddly enough, with Kempelen.

The careful replication of physiology was hardly restricted either to a few advanced or a few crackpot scientists, and was even, notoriously, taken as literally as possible by some inventors. Bell, himself, was judged an "oddball" by his Nova Scotia neighbors because word had gotten around that he had improved upon the *phonautographe* with "a device which used a human ear, procured for him from a friendly physician, where the bones formed the operating linkages." Later, during the rush to complete the telephone, he would, in the words of one biographer, experiment with dozens of mechanical forms "ranging from a three-foot diameter piece of boiler iron to one using the *actual bones of a human ear*. All worked."[51]

The experiments that led Bell directly to the telephone emerged from his research into the "harmonic telegraph," a device for transmitting multiple telegraph messages on a single wire. This device was directly inspired by insights derived from Helmholtz's "resonance theory" of hearing, which was itself based upon the precise analysis of anatomical structures. Helmholtz tested and "proved" his theory through experimental devices that duplicated in mechanical form the concrete physiological structures of the ear, following the path laid out by Kempelen. Ironically enough, Ernest Rutherford's challenge to Helmholtz's resonance theory, the "telephone" theory, took as its contrary explanatory model the very device Helmholtz had earlier inspired with his *competing* theory.[52] The two theories were then debated in the lab by means of a series of "mechalogues" or mechanical analogs of inner ear structures. Such models were supposed to demonstrate, by operating correctly, the validity of a particular theory. Thus, in theory or practice, all sensory researchers found themselves heirs, willing or not, to the great chess automaton.

As was often the case in more "mainstream" science, Kempelen, Helmholtz, and Bell's methods of "proof" involved the replication of natural phenomena. As several writers have argued, there is good reason to describe the principles followed by automata-makers like Kempelen and Jacques Vaucanson—maker of a renowned mechanical duck and an automaton flute player—as straightforwardly scientific in character.[53] (See figure 1.12.) In each case, the constructed device was expected to "prove" the scientific claims advanced by its constructor—if it worked, it could be considered to have demonstrated the point

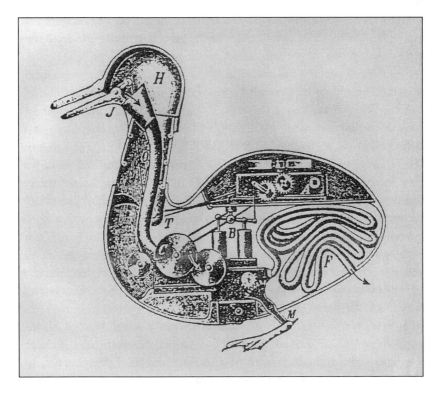

FIGURE 1.12
Vaucanson's duck.

in question.[54] Reaching far beyond the construction of automata, "mimetic experiments," which studied nature by literally duplicating the form of natural phenomena, comprised a major research tradition of the nineteenth century.[55]

Moreover, the imitation of nature, especially of the human body, found support in the belief of God's perfection, and specifically in arguments for God's existence from the design of the world, particularly in complex structures like the eye.[56] Since God was the perfect engineer, to duplicate his constructions would surely give one privileged access to the truth.[57] However, primary among the many factors conditioning attempts at sound representation was the simple fact that the object of representation was the human voice. Given the enormous cultural weight assigned to voice and language in definitions of humanity, it would have seemed only natural and even necessary to duplicate as closely as possible the organs that produced it. In contrast, the graphical tradition of sound representation arose from another direction and another set of concerns—ones that, at the start at least, had little to do with the voice.

In that other direction, but likewise bridging the gap between science and amusement, were those practitioners of the graphic tradition of sound representation. While the physiological study of vowels would continue far into the nineteenth century, the graphical method of analysis, with its emphasis on recording and visually examining vibrations, would soon come to dominate the field. Willis's insight that vowels ought to be studied independent of their origin like any other complex vibration, opened the doors to a new notion of sound representation that, like Edison's phonograph, cared little for the sound's source, seeking instead a permanent, scriptural trace of the phenomenon itself. Although Willis may be the actual historical source for this tradition, its more familiar origin is E. F. F. Chladni and his beautifully symmetrical *klangfiguren*.[58] Vibrating a dust-covered brass plate with a violin bow caused the dust to collect in patterned nodal points and produced these tone figures. The figures were widely recognized as having inaugurated a new era in sound representation because they produced semipermanent records of various vibrations through the agency of the phenomenon itself rather than the investigator. (See figure 1.13.)

Initiating a tradition that would find its greatest avatar in Marey, Chladni, like Robert Willis, Thomas Young, Jules-Antoine Lissajous, and Rudolph Koenig, sought a durable record of a moving, temporal phenomenon by enlisting the phenomenon itself as the motivating cause. Moreover, by creating "autographic" *visual* representations, these experimenters created a form that allowed easy comparison, analysis, and classification. The visual repre-

FIGURE I.I3
Chladni method of recording vibration, and Chladni plates.

sentation of sound, which could, like any other laboratory "inscription," be handled in familiar ways, quantified, retrieved at a moment's notice, and be analyzed regardless of the presence or absence of the source of vibration, suited the representational demands of the scientist far better than any physiological simulation could. Indeed, the more accurate the simulation, the less it differed from the phenomenon itself, and therefore, the less abstraction and analysis it permitted.

Although effective for certain purposes, physiological simulations of the voice could not produce permanent traces, and therefore they resisted repeated analysis and comparison. At the same time, they relied upon hearing, which was scientifically unreliable because of its inability to produce quantitative evaluations. On the other hand, visual inscriptions could be quantified, reproduced, circulated, and compared, as could, later, the traces inscribed on phonograph cylinders. Still, the notion of simulation did not disappear; it took on new functions. Instead of serving as the basis for the *production* of acoustic phenomena, the dominant deployment of this paradigm shifted to the duplication of hearing—the *receiver* of acoustic phenomena—and recording devices were modeled on the physiology of eyes and ears. As the epitome of sensitivity to phenomena visible and audible, the eye and the ear became paradigms for *graphical* instrument design, and devices designed in their image were developed to record light and sound.

The change in discursive and conceptual roles played by simulation— from simulations of the human sound production to the simulation of human perception—was accompanied by a very specific and correlated redeployment of scriptural tropes. While Edison's observations evoke an image of hearing as inhabited (both threatened and enabled) by writing understood as the nonpresent, the graphical tradition of scientific research emphasized writing as a form of mediated phenomenal presence, that is, as a trace of the very phenomenon it served to represent. The related ideas of self-writing sounds and of objects "imprinting themselves" on photographic plates nicely summarize the manner in which inscription as continuous bodily trace came to dominate some understandings of technological representation. Indeed, Scott's rather apposite use of the term "autograph" signals the manner in which writing might be imagined as a record of a unique phenomenal presence, against its more common understanding as an inadequate and unreliable substitute for speech. Writing, alternatively understood as a process of nonarbitrary and motivated inscription, linked the processes of representation to bodies and experience. At the same time, however, it was forever shadowed by the "problems" of writing signaled by Edison.

Together, the pair inscription/simulation served both to highlight these difficulties and new possibilities and simultaneously to suggest ways in which
they might be understood and assimilated.

A wide variety of phonographic devices were manufactured from actual
eardrums or other anatomical structures, and many others were designed to
mimic the ear, just as binocular cameras and viewers, imitating our pair of
eyes, appeared in the late 1800s. These obviously mimetic mechanical
"organs" testify to the extent to which representational technologies were
explicitly modeled on the perceptual faculties. By the same token, they also
suggest that perception itself was commensurate with these new media. If
one could, in some sense, duplicate perceptual capacities or even *improve* on
them by means of a mechanical device, then it stood to reason that the senses were understandable in terms of mechanics. By no means, however, did
these devices absolutely duplicate the senses since they lacked, among other
things, consciousness and reason. Yet they preserved *something* essential, and
in the case of Marey's graphical inscriptors, improved on those essentials.
The question then is precisely *what* was preserved, extended, or duplicated?
Moreover, we may ask in what ways these two apparatuses—mechanical and
sensory—mutually informed one another, or more specifically, how the senses came to shape the understanding of technology, and reciprocally, how
technology came to shape the imagination of perception.

In a manner of speaking, the reciprocal, or even dialectical, mediations of
these two terms—technology and sensory experience—are the very essence
of this book, for it is in and around these terms that all the major issues
revolve and toward which all of the investigations point. I suggest that to
open this can of worms we revisit Edison's essays on the phonograph, keeping in mind the importance of the pairs writing and simulation, perception
and representation.

Edison offers one interpretation of the defining characteristic linking photography, phonography, and cinema when he notes that the kinetoscope, like
the phonograph, effects "the recording and reproduction of things in
motion."[59] This somewhat curious locution implicitly understands the
phonograph to stand in the same relation to stillness as the kinetoscope does
to photography. However, rather than attend to conceptual difficulties in this
comparison, we might better ask what it sought to express. In his essays on
the phonograph, Edison returns again and again to the terms "captivation"
and "fugitive." He describes the phonograph's "founding principle" as "the
gathering up and retaining of sounds hitherto *fugitive*," and its accomplish-

ments as *"captivity* of all manner of sound waves heretofore designated as *'fugitive,'* " "their reproduction with all their original characteristics," "the transmission of such *captive* sounds," and "the *captivation* of sounds, with or without the knowledge or consent of the source of their origin."

Similar terminology informs Marey's descriptions of his graphical inscriptors, Nadar's discussions of photography and phonography, as well as countless other reports on these devices.[60] The idea of "capturing" phenomena that were previously fugitive or ephemeral likewise definitively shapes the theorization of scientific recording. Indeed, this is the most persistent manner in which Marey's devices are described. The possibility of preserving and repeating the previously evanescent and singular emerged as one of the most persistent themes associated with all manner of representational technologies and offers a rare opportunity to understand how tropes of writing and simulation served to negotiate institutional relationships to a new technological capacity.

The contrast between the fugitive and its various opposites (the stable, the permanent, the captured, the repeatable) has a long and instructive history in philosophy, science, and linguistics, where it is associated with ephemeral elements of a phenomenon that are unworthy of preservation since they escape the criteria of universality, durability, and permanence.[61] While this hierarchy might make sense to a linguist interested in avoiding undue emphasis on temporary usages, recent coinages, regional accents or vocabulary, it makes less sense to Scott, Bell, and Edison, who deploy linguistic tropes (from stenography to universal alphabets) in order precisely to elevate the fugitive and ephemeral to a prominence hitherto unimagined. By preserving the purely contingent, these phonographic systems effectively reversed the rational hierarchy between the essential and inessential, between substance and accident.

In the world of scientific experiment, particularly in the study of movement and vibration, such a reversal was necessary. While "fugitive" variations and contingencies of particular speech events hindered the project of the dictionary, in the field of physiology they were a godsend. Marey in particular made it a guiding principle that his instruments should become ever more sensitive, and their recorded traces ever more finely differentiated.[62] This increased sensitivity responded to the fact that, for physiologists, what had previously seemed "background" phenomena—complexities and oddities of movement—had suddenly assumed the foreground of their researches. In fact, as Marey repeatedly pointed out, it was precisely our expectations about what *should* be in the foreground that blinded us from the actualities of physiological complexity. Therefore, Marey's graphical inscriptors were capable of outstripping conscious perception because they were not only infinitely sensitive but also, in a sense, infinitely "passive," refusing the active hierarchies associated with habits of reason and expectations derived from theories not grounded in direct observation.[63]

In fact, these prosthetic senses soon served as models for the general attitude of the investigator himself. While devices like the *phonautographe*, the manometric flame, or any one of Marey's many graphical inscriptors outdid the human senses in sensitivity and storage capacity, they also suggested attitudes that could be emulated by human observers. Again and again, the mechanical senses' ability to capture the ephemeral and fugitive was attributed to their capacity to record all phenomena in an "unprejudiced" manner—that is, without regard to prior notions of intrinsic importance. As Lorraine Daston and Peter Galison have shown, in the sciences such perceptual "passivity," modeled on the photograph, emerged in the nineteenth century as the very paradigm for experimental objectivity.[64] It even became fairly common to describe human perception in mechanical terms, or as aspiring to the unbiased receptivity of a mechanical instrument. One typical instance argues that physicists must overcome the prejudice of the "musical" ear, and learn to give all aspects of a sound equal prominence or attention.

In *Pneumatics and Acoustics* (c. 1888), science writer John H. Pepper argues that hearing should play an increased role in scientific studies of sound. In a sketch of the techniques by which a student might effect an education of his hearing, Pepper distinguishes the scientific ear from the musical one—musicians offering the most widely disseminated model of educated listeners. Musical training, he argues, is so central to the idea of the educated ear that its norms have shaped not only the perceptual habits of scientists but the basic categories of scientific sound analysis as well.[65] Rather than valorize such training, however, Pepper notes, "We might even say that musicians are bad appreciators of sounds not used in music . . . [because of] a habit they have acquired of not listening to sounds which are grating to the ear. . . . The moment the sounds he hears no longer produce on him the impression of notes, he hears only noise, and can no longer judge of anything." He continues,

For the professor of physics, whose task is to appreciate sounds as a result of vibrations *however produced*, the case is quite different. Here there is no musical preoccupation. When we listen to a sound to learn something— and this is invariably our object—we may pay little attention to its distinctness. . . . What must be carefully listened to are the feeble sounds which always precede it, those which accompany it, and those which sometimes follow it. . . . Think of what immense resources the ear would procure us if, instead of hearing noises, it always heard sounds.[66]

Like the devices decisively altering laboratory practice around the world, scientists themselves were reimagined as ideally sensitive and ideally "passive" (here "objective") receivers of data. What these comments make clear by comparison is that like their human counterparts who had to work to unlearn categories of perceptual evaluation, "passive" mechanical devices effectively challenged or reversed established cultural and intellectual norms and hierarchies. In the course of this shift, previously "rational" and "objective" categories like "musical/nonmusical," "meaningful/nonmeaningful" or "essential/contingent" came to appear simply prejudicial. In essence, the very devices that initially had been modeled on human perceptual faculties had themselves come to embody a new paradigm of ideal and decidedly inhuman sensory acuity. In contrast to centuries of thought to the contrary, the new sensory attitude was passionately devoted to a counter-intellectual, counter-hierarchical sensitivity to the fugitive, the ephemeral, "the background."

While the resulting proliferation of observational detail recorded by the new devices offered scientists like Marey an unprecedented wealth of data from which to develop more precise analyses of human and animal physiology, it simultaneously presented a series of corresponding epistemological dilemmas. The strategic "passivity" of inscriptional devices, which refused to ignore those stimuli that contradicted expected results and which tirelessly and indiscriminately recorded all contingencies regardless of whether they corresponded to more general "types," effectively destroyed the reigning hierarchies of scientific observation and, even more so, representation. This is not to claim that such recordings did not in many cases confirm what observers had expected, but rather that what humans had long considered intrinsically important and unimportant features of phenomena might be presented with equal representational emphasis. Such a lack of inscriptional or pictorial hierarchy made the task of deciphering laboratory representations newly difficult. Without a priori guides to analysis, it was difficult and sometimes even impossible to read the resulting graphs. That is, too much "objectivity" might result in a wealth of confusing and irrelevant detail.

In the field of scientific illustration a correspondingly literal pictorial fidelity became an especially acute problem as artists and publishers sought to render images that would remain most responsibly "true to nature." Reversing decades of scientific judgment, nineteenth-century illustrators were increasingly expected to refrain from interpretive and generalizing techniques that sought to clarify images by selecting and emphasizing the "typical" in the individual case. A purely literal fidelity, however, did not always meet the demands of science either. Indeed, excessive fidelity was often criticized, as in the case of seventeenth-century anatomist Godfried Bidloo, who struck an 1886 observer as "too naturalistic both for art and science," and "almost Zolaesque in his superfluous realism."[67] While increasingly the paradigm of scientific fidelity, the sort of mechanical and subjective "passivity" characteristic of both devices and illustrators could create new representational dilemmas. One could even be "too faithful" to the extent that the ephemera or background details threatened to submerge the essential foreground and destroy the image's *legibility*. The same lack of subjective intervention that guaranteed fidelity by minimizing the impact of abstract, rational categories of perception and evaluation returned as a lack of intelligibility in the finished representation.[68]

Such was certainly the case with Marey's early attempts at analyzing human movement photographically. Photography had appeared as a nearly perfect recording medium for Marey's motion analyses, but it carried with it a number of interrelated difficulties. Automatizing and even embodying the ideal of indiscriminate and universal visual sensitivity, the photograph offered an image almost shockingly replete in information and detail. However, for Marey this was a mixed blessing. His early attempts to render animal and human motion resulted in a "Zolaesque" proliferation of detail that threatened, and at times destroyed, the legibility of the finished image. Recognizing that not every aspect of the movement under consideration merited equal emphasis, Marey ingeniously altered neither his photographic devices nor their vaunted sensory perfection, but the very subjects to be photographed, dressing them entirely in black and highlighting their anatomies with luminescent stripes designed to coincide with the major bones and muscle groups involved in a particular activity. By proceeding thus, Marey managed at the same time to take advantage of the camera's "passivity/objectivity" and lack of conventional pictorial prejudice, and simultaneously to eradicate those features that experimentation had shown to be irrelevant and confusing.

While the possibilities offered by these new "prosthetic" and "extended" senses were enormous and promising, automatizing perception rendered it both extraordinarily sensitive and extraordinarily indiscriminate. As the words *prosthetic* and *automatized* suggest, such senses, even when understood as "perfections," raised a group of problems that might, once again, question the role of "the human" in perception and representation. On the one hand, these devices possessed "superhuman" inscriptional and perceptual capacities, but on the other could produce "inhuman"—that is, unreasoned and irrationally illegible—images. Representations might abound in purely accidental details. Even the notion of "autographic" inscriptions, while promising ever more unmediated traces, seemed dangerously to evacuate human intelligence from the recording process by allowing phenomena to cause their own representations. In short, "perfecting" the senses in one direction required diminishing them along another axis of experience—one that crucially linked brute perceptions to human meanings through the mediations of reason. Redefining the relationship between representational norms and sensory experience in this manner consequently raised issues of a more general nature.

In the realm of photographic aesthetics questions of legibility and of irrationality take on a characteristic hue, coloring basic understandings of human intelligence and creativity in a context where the criterion of the "objective" carried little intrinsic weight. Here the relationship between perception—especially "artistic" perception—and representation served to anchor a variety of debates over photography and film (although, as I will argue later, they impinge upon phonography as well)[69] and suggested a limited but effective range of possible ways for dealing with images that corresponded but poorly to existing norms. In a world where the accidental, the purely contingent, the "too faithful," and the "passive" were increasingly a pictorial force—and certainly an incontrovertible presence—artists, poets, critics, philosophers, and image-makers of all sorts grappled with the basic questions the camera raised. If it were true that the camera embodied a more perfect eye, as much commentary suggested, surely it should also offer a more perfect model for picture-making. However, in practice, its superior sensitivity and its potential passivity were identified almost immediately as creating aesthetic *problems* whose "solution" involved rethinking not only picture-making but perception as well and, most importantly, the relationships between perception, representation, and meaning.

As in the sciences, the idea that an image might be *too* faithful fundamentally shaped discussions of photographic aesthetics. Writers even employed the same rhetoric, echoing a similar oscillation between a delight in the perfections wrought by the camera's inhuman and indiscriminate gaze, and consternation at its lack of respect for conventional hierarchies of importance. A critic surveying an 1895 amateur photography exhibition, for example, complained that he "could not help noticing the painful efforts of the photographers to include as much unnecessary detail in each photograph as Zola will put in one of his novels."[70] Many other writers reiterated these criticisms, claiming that photographs were too "graphically correct" and offered "too much infinitesimal and often meaningless detail."[71] Writers typically criticized visually dense images by comparison with "the great masterpieces of art, [where] we find that there is always a principal object or 'motif' to which everything else is subordinate."[72] Rather than the mastery and purposefulness characteristic of, say, Italian Renaissance imagery, such photos often exhibited a distressing lack of obvious human composition. Indeed, one might say that these images seemed to consider the subject (both maker and implied behold-er) *secondary* to the visible world and subservient to its vagaries.

However historically specific, this dispute was by no means entirely new. Debates over the coherence and unity of pictures have occupied artists and critics for centuries, and previous instances are illuminating in their choice of problems and issues. In this connection, consider this assessment of Northern European (Dutch) painting attributed to Michelangelo:

> In Flanders they paint with a view to external exactness or such things as may cheer you and of which you cannot speak ill. . . . They paint stuffs and masonry, the green grass of the fields, the shadow of trees, the rivers and bridges, which they call landscapes, with many figures on this side and many figures on that. And all this, though it pleases some persons, is done without *reason or art*, without symmetry or proportion, without skillful choice or boldness and, finally without substance or vigor.[73]

In other words, while still paintings, these images were not art. Indeed, one might reasonably say that they were not *pictures*.

As it happens, Michelangelo might just as well have been writing for the *American Amateur Photographer* in the 1890s. For photographic journals of the era, the very same distinction between image and "picture" (art) came to the fore as groups battled over the definitions of both photography and artistic picturing. In some measure, this dispute arose in response to the dissemina-tion of the Kodak hand camera, which greatly increased the number and vari-ety of photographic images circulating in the culture. In April 1895 one rep-resentative critic suggested that

> some years ago, when amateur photography was in its infancy here . . . a soulless corporation extensively advertised a camera which only required a button to be pressed and pictures were made. The idea soon took root that there was nothing in photography, when it merely required the pressing of a button. It was apparent that any fool could do that.[74]

In addition to the apparent lack of reason and skill involved in photography, the resulting images also lacked proper subject matter and composition. Photography's capacity or even propensity to reproduce anything and everything ("stuffs and masonry, the green grass of the fields, the shadow of trees, the rivers and bridges, which they call landscapes") and to render even insignificant objects "picturesque" tended to rob the image of the formal hierarchy necessary to be considered a "picture."[75] Moreover, by seeming to eliminate the need for conscious activity on the maker's part, Kodak photography appeared to remove all intellectual effort from the process of image-making and render the maker utterly passive in the face of the visible world and the machine.

Rather than serve as a concrete embodiment of the "innocent" eye, the camera threatened to become a mechanized and irrational one, producing pictures that, while "autographically" faithful, lacked any human meaning. As a surrogate visual organ, the camera's gains in sensitivity and thoroughness were offset by compensatory fears of intellectual, moral, and aesthetic mechanization. At its most radical, the camera made it possible, perhaps for the first time in history, to make a legible image *entirely by accident.*[76] This unprecedented possibility seriously questioned photography's value as art (indeed, as a coherent image) because it emphatically asserted the subject's utter lack of primacy in the process, implying that the reasoned hand of the artist, which for centuries had actively constructed pictorial worlds, might be abandoned for a model of perception and of representation that allowed the visible world to determine its own order, and let it wash over its passive eye.

Registering this fear, the "accidental" or unintentional character of many images occasioned great comment, and early descriptions of the daguerreotype repeatedly stress the unreasoned and arbitrary character of the resulting images, mentioning, for example, "the extraordinary minuteness of details," "the slightest accidental effects of the sun,"[77] and the "frightful amount of detail."[78] Critics similarly noted that "all the minutest [details] . . . were represented with the most mathematical accuracy,"[79] and that "every stone, every little perfection, or dilapidation, the most minute detail, *which, in an ordinary drawing, would merit no special attention,* becomes on a photograph, worthy of careful study."[80] Oliver Wendell Holmes reiterated this sentiment, arguing that the "distinctness of the lesser details of a building or a landscape often gives us incidental truths which interest us more than the central object of the picture."[81] Unlike a picture, then, a photograph appeared to be an "all-over" sort of representation—one that exhibits a single standard of treatment for all the objects represented—where seemingly insignificant objects are rendered in as much detail as human figures.[82] Here, the "more perfect" senses threatened to outstrip the intellect through their refusal to subordinate and normalize according to preexistent schemas of importance, resulting, as in the lab, in a confusing abundance of irrelevant detail.

Despite such concern, however, such images came to be valued by some artists and critics, and flourished in genres like stereophotography, which stressed a new kind of pictorial immediacy. The newly valued attitude of perceptual and representational "passivity" suggested here implied not only a rethinking of perception along lines suggested by the camera but also a reevaluation of the nature of visible reality. The visible world, rather than defined by stable forms and eternal relationships of harmony and order, is understood here as mobile, fluid, fugitive, evanescent.[83] In a parallel manner, perception in general, and aesthetic perception in particular, are vigorously reimagined in order to bring them into accord with newly possible forms of image production and forms of visual experience, and thereby reinstalled a rather traditional understanding of the relationship between perception and representation. The logic that valued these "incoherent" and "passive" images implied a different mode of picturing—one devoted not to creating classically organized images, but to recording the real perceptions of an artist immersed in the very world he sought to depict.

This mode was not without precedent, however, nor are its roots planted solely in the soil of aesthetics. The kind of interpretation I suggest—reading images as simulated perceptions—rose to prominence in a world where new forms of visual experience based on the positioning and repositioning of bodies in space proliferated in ever greater numbers. In the wider visual context of railroad and subway travel, panoramas, parades, stereograph lectures, snapshots, Ferris wheels, and scenic railways (all of which offered visual experiences whose character was determined by the spectator's literal location vis-à-vis a "changing" visual field), spectatorial "secondariness" or passivity was utterly familiar. One saw what one saw in these instances by being transported through space or because the visible world moved "past," beyond one's control. If a passenger on a train failed to notice an object as he hurtled through the landscape, he could not, at will, readjust his framing; his visual experience was determined by forces beyond his control, and at times contrary to his own sense of visual importance. In such visual worlds, one "caught" images as best one could, and arbitrary and contingent framings were the price to be paid.

These new forms of visual experience and the derivative sense of pictorial order that justified treating fragmentary and disorganized images as coherent pictures also found roots in the traditional visual arts. Jacques Aumont has pointed to a development within the history of painting he calls the "mobilization of the gaze," which describes a tendency, which he argues emerged between 1780 and 1820, in both painting and drawing, to destabilize the represented point of view.[84] Whatever the cause of their emergence, the paired tendencies that, first, represented objects previously considered in and of themselves unimportant, and, second, that used the frame as a sort of visual "scalpel," gained an unprecedented artistic legitimacy in these years. In a similar vein, Peter Galassi's influential essay, *Before Photography*, argues that the rise of such an aesthetic presents "a new and fundamentally modern pictorial syntax of immediate, synoptic perceptions and discontinuous, unexpected forms. It is the syntax of an art devoted to the singular and contingent rather than the universal and stable."[85] These are words that, no doubt, could be applied to many late nineteenth-century photographs as well.

Characteristic of this new disposition toward artistic practice was a reconceptualization of both the world-to-be-represented and the relationship between the artist, the spectator, and that world. In short, the landscape or, more broadly, the visual world, came to be imagined as a connected series of potential pictures that needed, principally, to be *framed* correctly. The very act of seeing, or of "framing," a preexistent world accrued an artistic value it had not previously possessed. Likewise, artists reconceived their own work as a recording of immediate sensory impressions directly in pictorial form. Aumont locates this shift in a change from the *ébauche*, or preparatory sketch for a painting, to the *étude*, which recorded an immediate impression not in view of future development or retouching, but as an act of direct representation itself.

Implicit in this transformation was the belief that vision and experience could be artistic phenomena in their own right and that pictorial order and significance did not necessarily derive from the intrinsic importance of the represented objects, nor from the historic or inherently important nature of the depicted act or event. The unpopulated landscape (which Michelangelo derided), the architectural fragment, the banal object, the narratively diffuse scene, or the scene whose vantage point makes its ostensible subject illegible, could acquire an aesthetic significance, provided it was understood to dramatize an experience of real seeing. The fragmentation of such images might be justified or motivated by the presumed presence of a looker whose *experience* was necessarily fragmented. Lack of pictorial hierarchy (the subordination of motifs) was reconceptualized not as a loss of mastery over the image, but as a function of the world *as seen*.

If neither the Coliseum nor the Basilica were readily identifiable in Corot's *View of the Coliseum Through the Arches of the Basilica of Constantine* (1825), this was because Corot and his implied beholder were part of the same physical world that these structures inhabited. They were therefore limited to concretely available, "actual" points of view rather than ideal and clarifying imaginary perspectives. If the image lacked the clarity of one of Piero's townscapes, this was because the real space offered no such "vantage point." Its material situatedness and intransigence limit our view of the scene. These are, to paraphrase Svetlana Alpers, not "objects viewed," in which an external viewer is presented with a view of an object, but rather "views of objects viewed."[86] Likewise, the fragmentary, illegible, or fleeting image made sense if one took oneself to be perceiving *with* the camera, as if the photo captured the pure contingencies of a visual world whose essential character was one of fugitivity.

As a corollary development, the importance of pictorial dramaturgy diminished, while arranging oneself in relation to an "untampered" world emerged as a fully "artistic" strategy—one that found its roots in the popularity of the "Claude Glass," a convex mirror with which tourists would "hunt" and "capture" landscapes.[87] Even the very act of picture-making was reimagined as a process of "capturing" recalcitrant and fugitive events. Metaphors of hunting and capture were common in the attendant literature of the picturesque, where the object was "to obtain a sudden glance, as [nature] flits past."[88] In short, these practices of vision presupposed an embodied beholder whose aesthetic activity assumed the priority of an untampered and unmanipulated world.

The importance of the beholder's physical location to picturing was recognized throughout visual culture, but perhaps most thoroughly and in the most practical way in stereophotography. In an 1859 guide to stereophotography, Thomas Starr King suggests that one familiarize oneself with an area in order to "learn not only just where the best *pictures are to be seen*, but also what a great difference is made in the effect of a landscape by a slight change of position on the road."[89] Clearly the creative act here is seeing—or more precisely positioning oneself in relation to the scene so that it "composes itself." The artist intervenes, therefore, only through "framing."

The equation of embodied looking with picture-making is not, however, confined to those in the business of pictures, for even Ralph Waldo Emerson had said, "To the attentive eye, each moment . . . has its own beauty, and in the same field it beholds, every hour, a picture which was never seen before and which shall never be seen again."[90] Although according to Emerson experience is best understood as transitory and fluid, each instant can be perceived as an organized image. Its uniqueness is a consequence of the fact that it exists in a flux *beyond our manipulation*, yet it is recognizable *as* unique, and as significant precisely through its being seen attentively, being *recognized* as a complete image. The exercise of a particular kind of vision—keen, attentive vision—transforms the banal into the significant or beautiful—into a picture. In essence, Emerson reimagines perception, metaphorically (and proleptically), on the model of the camera.

The pertinence of Emersonian aesthetics is far from coincidental, and its influence was not restricted to intellectuals. In a piece entitled, "Artistic Pictures: Suggestions How to Make Them," photographer and critic Elizabeth Flint Wade uses Emerson to support a *traditional* pictorial aesthetic by taking his "artistic eye" to be practicing a very conventional form of pictorial composition. Focusing on his belief that the eye can approach the flux of reality and "detach" an isolated, organized moment, the author offers a parallel with classical pictorial subordination: "Emerson says that the virtue of art lies in this detachment, in sequestering one object from the embarrassing variety, and in this power of detaching, and to magnify by detaching lies the quality and capacity of the artist."[91] Wade's use of the word "detach" eloquently captures one of the new image's more disruptive characteristics. So readily understandable was this view of the eye's relation to the visible world that Wade presents the following Emersonian description of artistic vision without attribution: "In Nature, the artistic eye finds a new and ever-varying aspect, even in the most familiar places."[92] Wade marshals these references in support of a theory of art that is based on "the great masterpieces," where compositional unity and subordination reign supreme.[93] The concept of artistic vision, as propounded by Emerson, however, assumes that such subordination (if in fact an "Emersonian" image would display subordination) is not a function of the artist per se, but of the flux of the world which *itself* forms organized images that the eye "detaches."

Significantly, Wade alters Emerson's "attentive eye," making it an "artistic" one instead. Like Pepper's listening physicist, the photographer's senses are asked to emulate a capacity exemplified, and even embodied, by a mechanical device for inscription. And while it would be rash to generalize from this comment and assert that all perceptual experience was irreversibly transformed by the advent of photography, phonography, and film, several more modest but still important conclusions might be suggested. In essence, each emerges from the confluence of concerns about perception, representation, and technology, and each sees the tropes of either inscription or human simulation as a strategy for coming to grips with new and sometimes confusing representational and experiential possibilities.

Both the scientific and aesthetic communities reacted to the emergence of the technical media with a mixture of excited anticipation and real concern over the changes they seemed to imply, if not always guarantee. As devices capable of autographic representation, the camera, the phonograph, and motion pictures offered dramatically expanded possibilities as tools of writing. Combining nonarbitrary and motivated inscription with the capacity to store previously evanescent phenomena, the new media superseded former modes of writing and promised to transform basic norms of representation in science and other fields. Simultaneously, they were understood to offer radically enhanced perceptual abilities—startlingly sensitive, indefatigable, and free from the prejudices and presupposition of human sensory habits. Like dozens of other machines in the nineteenth century, these devices seemed to offer ever-greater possibilities for automatized and prosthetic forms of human activity, extending and perfecting human perceptual capacities with unprecedented accuracy.[94]

Still, the same devices that promised a new and more accurate mode of writing and more perfect sensory potential were also prey to troubling objections associated with those very metaphors. While attempts to understand the new technologies as either writing or sensory simulation succeeded in describing new representational possibilities in familiar terms, they also served to signal familiar problems. As Edison indirectly realized, scriptural analogies may help us to grasp certain promising aspects of technological representation, but they also point to dilemmas traditionally associated with writing—its ability to be repeated out of original temporal and spatial context against the intentions of its author, its capacity for citation and reinscription in new chains of signification, and its propensity to signify in multiple, unexpected, and illimitable ways. Likewise, while the automatization of perception promised to expand sensory potentials, it also threatened to displace human reason, meaning, and creativity in arenas of practice and experience where previously they had been essential.

In a manner of speaking, the cases I describe from both the scientific and artistic communities hinge on those aspects of writing and of automatized perception that raise what I have called the specter of the "inhuman." Just as writing may divorce meaning from the presence and intention of its author, condemning it to circulate without anchor and in constant danger of illegitimate reuse, the photograph, the phonographic recording, and the film all similarly rendered visual or acoustic events material, iterable, and citable. Hence, as Edison noted, the technical media offered startlingly new potentials for surveillance. Such concerns and possibilities while describable in terms of writing could not be divorced from these media's apparent relationship to human perceptual experience.

While traditional forms of writing might be said to "store" only other written texts or, at best, abstractions of spoken language, the new media seemed to store *experiences*. In effect, these technologies seemed capable of offering new forms of perceptual experience that, while mediated, appeared to preserve the full phenomenality of an event. Like a telescope or a telephone, it seemed, photography, phonography, and film simply extended a perceptual experience across space and time without significant alteration— or with negligible alterations of the sort produced by mirrors and eyeglasses.[95] Nevertheless, these experiences were "inhabited" by writing in the sense that, even if understood as ontologically identical to "normal" perceptions, they could be cited and repeated, linked to new contexts, and forced to signify in ways contradictory to their original meanings. In short, they confronted the world with a form of *experience as written*—as enabled and endangered by the possibilities of writing as described in philosophy.

Each trope for understanding representational technologies—as a form of writing or as a simulation of human functioning—identified important cultural, semiotic, or aesthetic problems and translated them into a reasonably familiar form. If cultures had been able to deal with the vagaries of written language over centuries of history, then presumably they would be able to deal with these new forms of inscription. Similarly, if the new media bore an understandable relationship to sensory experience, then they could be dealt with in familiar ways. However, since in different ways metaphors of writing and prosthetic experience suggested that the inhuman had insinuated itself into the very heart of the human, and what is more, *combined* the "inhuman" qualities of both writing and automatization to create new representational crises, these familiar tropes were estranged. Therefore, when Marey's photographs exhibit an illegible proliferation of detail, or when "artistic" photographs appeared similarly bereft of meaning or coherence, it was precisely the inhuman—lack of subjectivity—that served as the locus of the problem.

In both instances, participants combined discursive and practical strategies to recuperate these inhuman potentials within an expanded understanding of the human. In the case of science, the inhuman was recuperated as the objective, while in aesthetics it was recuperated by valorizing a form of passive perception that allowed the world its own order, intervening only to "capture" it in its natural flux. Both institutions, in terms appropriate to their own representational demands, practices, and habits, confront the dilemma of the automatic or inhuman through a careful *redeployment of the subject.* That is, where certain aspects of technological representation threatened the coherence of human understandings of perception, representation, or aesthetics, commentators, practitioners, and critics found ways to "reanthropomorphize" them, and thereby render them intelligible and familiar.

In each such case, an initially threatening attribute of a representational technology—lack of representational bias, extreme sensitivity, tirelessness, repeatability—ultimately underwrites a new mode of institutional practice, while at the same time somehow threatening the stability of other established practices. In each case as well, a technological capacity, which in some way seemed inhuman, is remotivated to serve as the ground for a new understanding of human experience. In the laboratory, for example, the autographic or nonsubjective mode of inscription became the model for a new experimental "morality" of nonintervention, and thereby served to rehumanize and rationalize potentially irrational phenomena.[96] Simultaneously, mechanical possibilities served as a foundation upon which to redescribe and reimagine human faculties like perception and memory. It is hardly coincidental that, during this period, the words "photographic" and "phonographic" were first deployed as metaphors for uncannily precise and even omniscient forms of observation, representation, and memory.[97]

In parallel fashion, the camera's "inhuman" lack of representational intelligence, and its similarly inhuman capacity for capturing unassimilable detail, were recuperated by means of a new aesthetic devoted to "immediate, synoptic perceptions and discontinuous, unexpected forms,"[98] based in a revaluation of perceptual acts (i.e., seeing) as themselves aesthetic. Emerging from a wide variety of new visual experiences, each of which produced a visual display predicated on a "passive" or "subordinate" relation to an ever-changing visual field, this mode understood pictures to be simulated or captured perceptions. Their pictorial contingencies or incoherences simply replicated our fleeting and imperfect experience of the world. Since perception, however, was not typically imagined to be "passive" in this sense, experience itself was redescribed in terms suiting the properties of the camera and the phonograph.

The mediations between technology, perception, representation, and institution served to normalize unfamiliar forms and new possibilities by assimilating them to altered understandings of the status quo. In a process that has been recapitulated dozens of times since, an American culture on the threshold of modernity negotiated a new (and by no means untroubled) relationship to experience and experiential possibility. Indeed, it is the basic structure of Hollywood's reaction to the possibilities of recorded sound. In the next chapter, I will address this process of accommodation, negotiation, and assimilation in more detail, with an eye toward different and more difficult issues. Put bluntly, I will examine the consequences for day-to-day representational epistemologies when confronted with technologically mediated forms of perceptual experience. Once again, the tropes of writing and simulation will serve as touchstones since, in a way, I wish to examine what happens to the notion of *reference* when perception may be thought of as fundamentally inhabited by writing—that is, when it no longer obeys a logic of presence. What happens to ideas of sensory immediacy, accuracy, and authenticity when confronted with the evidence of mechanical senses—photographs and phonograms—that present themselves as simultaneously more accurate and yet completely inhuman in range and sensitivity? Can we trust the "vision" of a camera that "accidentally" witnesses a murder?

CHAPTER 2

PERFORMANCE, INSCRIPTION, DIEGESIS
The Technological Transformation of Representational Causality

Despite the proliferation of commentary on the century's newest devices for making pictures and recording sounds, the emergence of new representational technologies was, of course, not simply a matter of public discussion and debate. To concentrate on what was *said* about devices at the expense of the devices themselves risks minimizing the materiality of these forms while it simultaneously risks overemphasizing (to the point of idealism) the role of discourse. However, neither discourse nor device is as neatly definable as these warnings might suggest.

As I do elsewhere, here I assume the importance of proximate verbal discourses to an adequate understanding of the technical media, in part, simply because they provide some of the few available traces of otherwise ephemeral phenomena like practices of representation and reception. But beyond questions of empirical verifiability lies another more important concern, one that goes to the heart of my entire project. Our understandings of technology and of technological representation need to be revised radically. On the one hand, we need to recognize that in spite of their apparent material intransigence, representational devices, however familiar, are neither static nor self-defining. Put more forcefully, we need to take seriously the notion that, for example, a camera in one cultural/practical/representational context is not the "same" technology as the identical device in another, and that these differences are not trivial. Even sophisticated arguments like Jean-Louis Baudry's and Jean-Louis Comolli's about the aesthetic or social ideologies

"inherent" in photography or, more commonly, in cameras and lenses, necessarily reduce the heterogeneous field covered by those terms by naturaliz-
ing one particular historical tradition of representation as the exclusively
determinant one.[1] Instead, we should understand devices as constitutively situated in networks of assumptions, habits, practices, and modes of representation that extend well beyond instrument-centered definitions of technology. Likewise, it will be important to recognize that scientists, technicians,
instrument-makers, and engineers are social, political, and aesthetic actors as
well as technical ones—actors whose particular expertise and institutional
authority enabled their decisions, standards, and assumptions to determine
aesthetic and experiential form for broad segments of the population at large.

Therefore, throughout this study it will be evident that not only are technological form and development intimately related to reigning aesthetic
norms, they are at times in conflict with them as well. At particular moments
a misfit between instrument design and representational demand results in
the obsolescence of a device, and at others in a reimagining of representational technique. The dialectic between instrument and aesthetic/practical
context is therefore crucial for understanding how technologically based
media like the cinema develop over time. While there is not necessarily an
isomorphic or exclusive relationship between a particular "theory" or aesthetic and a technological form, neither is the relation arbitrary or negligible.
It might therefore be important to recall that on several occasions Gaston
Bachelard described scientific instruments as "reified theorems," or as others
would have it, "ideas made brass."[2] What Bachelard claims about scientific
instruments can serve as a heuristic for the study of representational technologies more generally. Rather than simply assume the truth of his claim,
however, we need to interrogate the complex historical process by which
such reifications take place, or how technical possibilities come to *appear* as
reifications of a particular theory. In short, one of the goals of this chapter
will be to shift historical attention from the purely technological to the discursive and practical, from a succession of devices to the ongoing struggles
to define and control the impact of these new media. Ultimately, I hope to
replace apparatus theory's insistence on the "built-in" ideology of the cinematic technologies in favor of an approach that sees contradiction between
levels of mediation, and that emphasizes the heterogeneous nature of technology and the local and contingent nature of ideological determination.

Stressing the importance of discourse and practice to the historical definition of representational technologies will enable us to estrange these too-
familiar devices and recapture the heterogeneous and often contradictory

possibilities they initially offered, and perhaps still offer. Among the first steps in such a venture is to rid ourselves of a few pervasive but misleading assumptions. The most disabling of these is perhaps the most common and one that concretizes several strands of my analysis—a reification embodied in the camera's "click." The apparent punctuality and unity of this event disguise a more complex and disjunct series of technologies and practices that *together* define the photographic process. The single, almost instantaneous exposure of a negative by means of a tiny and unique gesture encourages us to regard both the device and the act as similarly unitary. Nevertheless, a moment's reflection will confirm that even the most basic photograph involves a series or a "family" of technologies both photochemical and optical—technologies of detection, depiction, and reproduction.[3]

Indeed, we could subdivide the photographic "act" even further into more numerous or different processes, but regardless of our specific categories, the act of subdivision itself is a step in the right direction. Opening up our understanding of an instrument—and as we shall see, a representational act as well—allows us to understand how a particular representational technology is constitutively situated within different institutional and practical settings and how each calls upon photography's multiple capacities in significantly different ways.[4]

Various historical configurations of technological capacity, representational practice, and institutional necessity have defined given representational technologies in profoundly different ways—ways that deserve to be recalled if we wish to understand the complex and far-reaching manner in which these devices transformed life in the modern era. In the process of coming to grips with these transformations, both technological discourses and representational institutions serve as crucial arenas within which groups negotiated the significance, impact, and scope of new technologies. Crucial to these processes were the strategies by which knowledge, authority, and power came to be distributed among the members of a particular community and how these were deployed to maintain or alter established representational norms. Thus, the differential distribution of professional expertise and authority was fraught with social, cultural, and political ramifications of the most pressing sort.

In the long arc describing the historical trajectory of technological representation, entrenched habits of understanding and interpretation often ran headlong into new technical possibilities that seemed to threaten their legitimacy, yet contrary to both the most utopian and most pessimistic forecasts, change happened only with resistance, in discontinuous leaps, and by strategic productions and disseminations of knowledge. However, neither every discourse nor every discursive shift is as significant as every other—in fact, some are virtually meaningless. If we are to avoid the ever-present pitfall of a historical idealism that remains situated purely at the level of discourse, we must determine the manner by which particular discourses impinge upon practices and individuals. If, as I contend, discourse, practice, and institution are as important as device to an adequate understanding of representational technologies, we need to examine the manner in which they are mediately brought to bear on the conduct of everyday representational life.

The most often-told story of the cinema's emergence describes the debut of the Lumière brothers' *cinématographe* at the Grand Café where a naive and terrified audience flees an approaching train. For better or worse, this motif has come to embody the cinema's primal scene, serving as a metonymic touchstone for the history of the cinematic spectator, and suggesting a set of issues and a course of research centered on questions of belief, immediacy, and illusion. So profound is its capacity to concretize some sort of fantasy about the cinema that the story has persisted in all manner of accounts despite a general lack of evidence that any such violent response occurred on that December evening in 1895.[5]

While Tom Gunning provides an insightful and compelling argument that this story and others like it register the role that astonishment rather than credulity played in cinema's early years, he does not, in the end, exhaust either their variety or their persistence.[6] Why, after all, would such an aesthetic be described in terms of hyperbolic belief rather than, say, its peculiar rhythms of narrative and spectacle? And, if these stories only registered the role of astonishment, why would they be accompanied by other variations where astonishment seems to play no part? Rather than assume that such stories describe in some mediate way the responses of actual audiences, I propose an alternative approach that is both more sensitive to their specific texture and more suspect of their evidential status. I hope to show that the dozens of stories describing spectator response in terms of "realism," immediacy, illusion, and credulity relied on the systematic variation and obsessive reiteration of only a handful of tropes and that while these tropes may tell us little about the cinema per se, they tell us a great deal about the epistemological and representational dilemmas made palpable by the emergence of the technical media in general.

As Gunning suggests, these stories do not tell us much about whether audiences believed the images they saw on screen to be real, but they do tell us other things, indirectly. In their variety and repetition, these stories attempt to grasp the peculiar and novel relationship the film image established between its audience and its represented worlds. They sort out the medium's myriad representational and referential possibilities in a kind of vernacular "theory" that not only de-essentialized film (multiplying both its "ontologies" and its effects) but suggested a set of practical and conceptual strategies for understanding the nature of the new medium. In essence, these stories replace an ontology of *recording* with a pragmatics of *representation*. And while stories of terrified audiences are perhaps the most memorable and compelling, they merely begin the process by which a broader public works from the fact of the film image "back" to the world it represents, discovering, in the process, the possibilities and limits of cinematic representation.

That an improbable story of preternatural immediacy should stand metonymically for the cinema as a whole is, from another point of view, hardly surprising. Like Leonardo's own Renaissance rubes, whose inability to distinguish between painting and reality serves as an argument for the superiority of Venetian artistry, the early cinema's overly credulous and somehow less-than-fully human audiences seem constructed to make a point about something other than the actualities of early film spectatorship. The cast of nineteenth-century characters who flee turn-of-the-century screens are, with the exception of the Lumière example, all too familiar: rubes, drunks, and beasts. That these stories register fears about the suggestibility of the masses, as opposed to an urban middle class to whom they were presumably addressed, appears incontestable. Given the last century-and-a-half's concern with crowds, public representations, and persuasiveness, it would be difficult to ignore the extent to which such accounts serve as cautionary tales about the apparently uncanny immediacy and ideological effectiveness of the cinema. Surely at some level, too, they register class-based anxieties about the growing power of the working classes and signal the threat implied by the rise of mass audiences and mass spectacles.[7]

To understand these stories as allegories of ideological manipulation and middle-class fear is, when all is said and done, simply too crude. While questions of immediacy and persuasiveness must be, in any interpretation, central to their significance, the accounts themselves militate against such familiar readings. On the one hand, research shows us that early audiences were uniquely sophisticated in their reception practices, and that, regardless of class, responses were multiple, mixed, and highly contradictory.[8] On the other, stories where spectators flee an image on the screen are only one part of the complete anecdotal picture. In fact, at least two other types compete with these stories for prominence, neither of them implying the same class dynamics—in fact, here the protagonists seem implicitly bourgeois.

As the case of Leonardo makes clear, stories describing the uncannily real-istic character of images have existed for centuries, but they nevertheless burst forth in unprecedented number following the emergence of the cinema in the mid-1890s. The sheer volume and regularity of these accounts suggest a culture coming to grips with a novel, compelling, and threatening phe-nomenon—something that fundamentally engaged people's sense of the real, or more specifically, the boundary between the real and the represented. What is more, the reiteration of such tales and correlated expressions of per-plexity suggest that existing representational and experiential categories offered little purchase on the new phenomena. Time and time again cinema trade papers and contemporary periodicals describe staged robberies or drownings that provoke the intervention of passersby, films whose strong sense of immediacy force one to share the photographer's experience, or ones whose diegetic illusion is so convincing that enraged and enraptured spectators leap through the screen, attempting, as it were, to peck at Zeuxis's grapes.

Such stories, I would argue, served normative and pedagogical rather than descriptive ends, on the one hand negotiating and describing the significance of new representational technologies, and on the other ordering the fields of production and reception so that these unruly forms might better conform to existing norms of understanding. In fact, stories of the cinema's uncanny immediacy—its ability to produce a compelling, if imaginary, world—share certain key concerns and motifs with those attending the emergence of pho-tography and phonography and suggest that rather than limited to film and vision, they addressed concerns appropriate to all technologically mediated forms of perceptual experience. The cinema's concern with uncanny imme-diacy is neither typical of photography, nor of phonography, however, and each technology grappled with the problems of technological representation in its own manner.

Photography, phonography, and film elicited such notable and emotional response because they helped alter the existing balance between the senses, representation, and experience. These simultaneously compelling and threat-ening media did not simply offer new representational possibilities but also brought troubling epistemological issues to the fore in a dramatic and star-tling way. In simplest terms, their persuasive representational effects suggest-ed a referential guarantee that was perhaps not warranted by the actual cir-cumstances of their production. They seemed to offer the world in all its immediacy, yet audiences were simultaneously aware that they might be per-suasive fictions, since they contradicted accepted understandings of representational process. As I argued in chapter 1, representational technologies were routinely understood on the model of the human senses, but while the evidence of these senses seemed just as phenomenally present and rich, it seemed quite a bit less reliable than their flesh-and-blood counterparts. Here, it seemed, were recorded sense impressions that nevertheless could not guarantee their own veracity, since they did not truly put one in the presence of the phenomenon in question, but rather offered an account of it, a "quotation," if you will. That very representational or referential instability demanded that the cinema be recognized not as a single thing nor as implying a single relationship between representation and represented.

Public discourse (fictional, philosophical, legal, anecdotal) about these media played a crucial role in sorting out their scope and meaning, and although my ultimate aim is to explain phonography, I shall begin with discussions of film and photography. I do so not to imply a simple equivalence between the visible and the audible, but because controversies around the visual media make certain issues relevant to the phonograph clearer, and do so in terms that are familiar. I also wish to argue that parallel histories of the three media can highlight issues of a more general nature spurred by the possibility of technological recording, whatever the form.

ASTONISHMENT AND EXPLANATION

Stories about the early cinema's uncanny presence take their place within a wide variety of discourses that served to mediate between new sensory technologies and a lay public. The popular reception of nearly every nineteenth-century representational technology was shaped by one form of theatricalized presentation or another, often including lectures that both spectacularized and explained novel media. Charles Musser, for example, suggests that we locate the emergence of cinema within a dynamic tradition of "screen practice" incorporating all forms of projected entertainment. Arguing against diffuse cultural histories based upon "borrowings" and "influences," he insists that cinema's specificity demands a narrower but richer focus.[9] However, Musser's invocation of seventeenth-century Jesuit Athanasius Kircher's magic lantern presentations as a prototypical starting point for this analysis seems as misleading from one point of view as it is appropriate from another because its narrow focus limits our view of the relevant public context.[10]

Kircher's technique of combining dumbfounding visual illusion with demystifying explanation surely anticipates the conditions of much early cinematic exhibition, but this basic strategy was by no means restricted to the screen. Such tendencies were widespread in the worlds of amusement and education and, in fact, extended well beyond Kircher's lifetime and far into other disciplines, including, for example, acoustics, Kircher's primary area of expertise.[11] Indeed, performances combining astounding display with edifying commentary were deeply enmeshed in a more general transformation of knowledge and education with roots in the overlapping worlds of science and entertainment. Thus, while Musser's arguments about the demystification of the screen are accurate, the broader presentational structure he describes applies more generally to the relationship established in the eighteenth and nineteenth centuries between audiences and other sensual technologies.

In a parallel fashion, Barbara Stafford has argued that Enlightenment publics exhibited a distinctive passion for "rational recreation," that is, pleasing and effective forms of instruction that allowed a "legitimate indulging of the eyes in nondelusory patterns and mind-building shapes."[12] The popularity of these forms derived in part from their reliance on novel and sensuous technology, but these very devices were the province of the quackish faker as well. Thus figures like Étienne-Gaspard Robert (aka Robertson), the self-proclaimed inventor of the phantasmagoria, strenuously sought to differentiate his own techniques from those that skeptics might likewise have associated with the mountebank.

In his 1831 memoirs, for example, Robertson "takes pains to separate superstitious sorcery and prodigies from his own rational trials, supposedly presented 'without charlatanism.' "[13] He claimed to use phantasmagoria in the form of an "experimental trial" yielding not mystification but "group illumination," and by situating his own work within the tradition of scientific education, he avoided both social and ecclesiastical censure.[14] Likewise, the many similar lantern shows that flourished well into the American 1840s routinely included technological explication as part of the amusement.[15] Systematic demystification played an important role in legitimating these devices, both as inoffensive forms of entertainment and as technologies capable of making particular epistemological claims. Musser has further argued that demystifying the magic lantern—that is, purposely removing its effects from the realm of "actual" magic—provided a necessary foundation for the later emergence of the cinematic spectator because it consequently located screen images and effects firmly within the realm of "art."[16]

Photography, phonography, and motion pictures encountered their publics under similar conditions—through the mediations of either a flesh-and-blood lecturer-showman or the various written and oral discourses in cir-

culation around the new media. Therefore, when I suggest that widely disseminated anecdotes about these media served a quasi-pedagogical purpose,
I understand them as belonging to a residual tradition of this sort. That is, as
compelling—even mesmerizing—forms of sensual display, these technologies courted their own charges of fakery, charlatanism, sorcery, and mysti-
cism just as surely as any eighteenth-century experimentalist, and early proponents similarly responded with a combination of sober edification leav-
ened by spectacular amusement. However, rather than simply import a
basically European explanatory framework to the United States, I would
agree with Musser that the combination of astonishment and explanation
found its American form both as part of an "operational aesthetic" that Neil
Harris has identified as characteristic of nineteenth-century American culture, and in the context of a marketplace where these strategies served to
spur popular interest and increased consumption.

Like their European counterparts, American purveyors of the operational
aesthetic combined amusement and astonishment with detailed technical
explications, demonstrations of process, and opportunities for exercising critical judgment.[17] P. T. Barnum's exemplary success, Harris argues, rested in
part on flattering the characteristic skepticism and technological sophistication of American audiences by inviting them to "debunk" mechanical and
natural wonders like the infamous "mermaid" concocted from the corpses of
a monkey and a large fish. (For the privilege of debunking, of course, one still
had to buy a ticket.) In the case of traveling lantern and phonograph
exhibitors, elaborate educational lectures about the pictured or recorded
objects were combined with an explanatory discourse on the new technologies. These showmen combined a middle-class interest in education and
"high-class"[18] entertainments with a more pervasive interest in the demystification of astounding technical marvels like the telescope, telegraph, and
phonograph. So neatly did photography fit this general pattern of astonishment and skepticism that early reports of the daguerreotype process were
dismissed by a prominent American chemist as no more than a reprise of the
infamous "moon hoax" of 1835, when the New York Sun reported that Sir John
Herschel had taken a new telescope to the Cape of Good Hope and seen
actual moon creatures.[19] Such skepticism about technological capacity was
perhaps matched only by its obverse, technological utopianism.

If in Europe the vogue for scientific demonstration lectures began among
the aristocracy and later achieved great success with the middle classes, in the
United States popular interest in science took a characteristically more "democratic" (and commodified) form, aided by publicly funded education and

enormous markets for novel gadgets. In the wider social context, the dissemination of technological and/or scientific knowledge by way of amusing experiment and demonstration found other forms of support. A wide variety of popular books sought to teach the principles of science through compelling but simple experimentation and explanation. While, for example, David Brewster's immensely popular *Letters on Natural Magic* ostensibly sought to demystify the various optical and acoustic illusions perpetrated by hoaxers and pseudoscientists, others adopted the formula in order to renovate American scientific education or to promote science as a form of popular amusement.[20]

These tendencies were reinforced by the soaring popularity of recreational literature devoted to technical and scientific explanation and the great success of magazines like *Popular Science Monthly* and *Scientific American*, which regularly published essays devoted to new technologies. To these were added the promotional and pedagogical literature distributed by stereograph, phonograph, and camera companies, whose profits depended in no small measure on technical demystification, or more accurately, the regulated dissemination of specific knowledge about the technology.[21] Taken together, newspapers, popular magazines and literature, promotional materials, popular science texts, lecture demonstrations, popular songs, melodrama, hoaxers, itinerant showmen, and museum displays formed a network of institutions and practices devoted to discussing, describing, and explaining technological novelties. Therefore, technical explanation found well-tilled and fertile ground upon which to flourish, and soon formed an integral part of the American economy of amusement.

Clearly, while providing genuine demystifications of both natural phenomena and potentially confounding technological displays, neither the proliferation of such strategies nor their prominence with regard to new representational technologies in particular can adequately be explained by a simple desire for knowledge. Concerns of a more immediately visceral sort seem to have spurred the surprising range of commentary on these novel devices, and it became not only common but, I would argue, expected that explanatory or pedagogical discourses would situate and define new technologies as a matter of course.

Vernacular discussions, humorous stories, gossip, and imaginative speculation about new devices formed one of the most pervasive and effective forms of technological familiarization. By offering a simple taxonomy of such explanatory discourses, beginning with those adjoining the cinema, I hope to demonstrate how each story type constructs both a different notion of "realism" and a different understanding of the filmic sign. In fact, "realism" (which is a notoriously slippery term) is finally not at stake in these stories. What goes under the cover of "realism" or "immediacy" is neither, but rather an analysis of the possible causal circumstances of technological representation. To early audiences, filmic and photographic images seemed to bear a novel and special relationship to what they depicted—genetically distinct from other forms of imagery—and "realism" served as an umbrella concept under which to grasp that relationship and to order the multiple processes of technological representation. By segregating opposed, and sometimes contradictory, sign functions within different realms of representational production and reception, these stories bring conceptual and cultural order to the unfamiliar and often threatening forms brought forth by the new media.

TAXONOMIES: BEYOND THE GRAND CAFÉ

While the most commonly discussed cinema anecdotes are of the "Grand Café" type, there exist at least two alternate story types. The first are those devoted to gullible passersby who are so convinced of a *staged event's* authenticity that they intervene in the action. For example, in 1907 the *Moving Picture World* tells us that five hundred "trembling onlookers" waited breathlessly in Central Park for the "rescue" of a drowning mannequin, and that a New Jersey undertaker foiled an ersatz bank robbery by drawing his gun. We similarly learn that a feigned drowning at Atlantic City caused "panic stricken" onlookers to run desperately up and down the beach, and that the film's hero had to compete with a throng of unexpected rivals in order to effect his rescue. Other accounts tell of two Newark citizens who disrupt a staged horsewhipping, and that even the police are sometimes fooled by staged crimes.[22] The *Views and Film Index* from that year likewise offers a large number of such tales.[23] Contrary to the broader Grand Café paradigm, the gullible here are by no means ignorant or uncultured rubes but solidly middle-class or even professional citizens, suggesting that these stories are not simply designed to flatter the knowledge of the informed reader (although they do that, too). Instead, these stories focus our attention on the persuasive *staging* of an event, highlighting its potential role in the processes of cinematic representation.

The second story type makes the claim that particular images derive their immediacy either from the very process of technological inscription or by forcing a sort of perceptual identification with the cameraman. Early press reports on the cinema routinely bore titles like "Moving Pictures and the

Machines That Create Them," and "Revelations of the Camera," indicating that the device itself assumed an agency that both reproduced the realm of appearances and altered the relations between the visible and invisible by revealing what had previously been suprasensible.[24] The camera operator similarly came to the fore in discussions of filmic panoramas where we are told that we can see a scene "just as truly as the Bioscope operator who took the picture," and when a distributor "present[s] here a series of interesting pictures and show[s] a number of scenes just as witnessed by a visitor."[25] In fact, an entire subindustry of stereoscopic photography was premised on this very notion of experiential surrogacy, and company literature explicitly promoted this sort of spectatorial engagement.[26]

Crucially, both of these story types center their understanding of immediacy in the realm of image *production* rather than reception, linking verisimilitude to the duplication of profilmic and filmic actualities. By contrast, the familiar "Café" type includes those stories that, *regardless* of the actualities of production, present a convincing illusion when shown on screen. In these a dog makes a mad rush for projected ducks, a "partly intoxicated" man shoots five times at a villain, a fox terrier attacks a pictured rat, spectators warn characters of impending danger, two intoxicated Oklahoma City men shoot at a Russian bear, still another dog races a screen rival for a stick while one more attacks a sheep in a Passion play and yet another plunges into the screen to save a drowning boy.[27] In these cases, the plausibility of the represented world is diegetic rather than literal, that is, purely a function of a projected image whose spatiotemporal specifics might bear no strict relationship to those in any profilmic or filmic actuality.

In addition to the manifest level concerned with illustrating three ways of overengaging the unsuspecting, these stories also suggest three distinct arenas for representational manipulation—a realm of performance, one of inscription, and one of diegetic construction. By highlighting these three zones of intervention, the stories mark out distinct parameters of cinematic technique. Intervening in each arena, the stories tell us, will result in different representational effects and simultaneously raises different epistemological or semiotic issues. Depending on how a particular film is manipulated, it might or might not be an accurate record of some real event, just as it might or might not present a view of a real space and time. While specific to the cinema, these divisions find suggestive parallels in the comparable discourses surrounding both photography and the phonograph, where similar basic questions of representation and reference arise. While not wishing to overstress the similarities between distinct media, I would suggest that the tentative parallels between them allow us to understand something unique and fundamental about technological forms of representation.

NATURE'S PENCIL: AUTOGRAPHY AND REPRESENTATION

While early responses to the cinema described its novelty as its ability to confuse representation and reality, early discussions of photography centered, instead, on issues of pictorial agency, asking in effect what "caused" an image to take the particular form it had. As I noted in the previous chapter, the earliest attempts to describe these new pictures located causal primacy in the very objects photographed or, more generally, in "Nature." Daguerre, for instance, described his invention as the "chemical and physical process which gives Nature the ability to reproduce herself," Oliver Wendell Holmes wrote "the honest sunshine is Nature's sternest painter, yet the best," and Lady Elizabeth Eastlake that "light is made to portray [the object]." Photography was described as images "drawn by sunlight itself," as a process that "gives her [Nature] the power to reproduce herself," and as "nature copying nature, by nature's hand." Most famously, Talbot subtitled his paper to the Royal Society, "The Process by Which Natural Objects May be Made to Delineate Themselves without the Aid of the Artist's Pencil."[28]

An "autographic" understanding of representation characterized early responses to phonography as well. Nadar, in fact, imagined the two media as essentially related along these lines when he forecasted the possibility of an "acoustic daguerreotype." He may even have thought his premonition realized five years later, in 1861, when Edouard-Léon Scott read his papers concerning sonic representation before the French Académie des Sciences.[29] Like the daguerreotype, the phonograph immediately elicited dreams of automatized representation, of objects drawing their own pictures. The very name of Scott's recording device—*phonautographe*—bears witness to this aspiration, and nearly every account of its invention employed the language of "nature's own script" or "sounds that write themselves."[30] This rhetoric continued a tradition begun decades earlier in response to phenomena like Chladni's sound figures (*klangfiguren*) and Marey's attempts to render biological movements, as he put it, "in the language of the phenomena themselves."[31] Indeed, it became the goal of scientific recording in general to deploy the phenomenon under analysis as the agency of its own documentation.

Across media and across institutions, the discourse of object-causality underwrote claims to unprecedented accuracy and dreams of scientific and aesthetic perfection precisely by removing human agency from the representational process. After all, what could produce a truer or more beautiful image than nature's own hand? Simultaneously, however, parallel and often competing discourses centering on the devices, on the operators, and on the finished representation pure and simple rivaled Nature as the determining factor in representational persuasiveness.

As was the case with early cinema, some stories or accounts concentrated on the processes of inscription, asserting either the operator or the machine itself as the agent. Here photography provides a more illuminating case than does the phonograph. P. H. Emerson is among the most notable critics to argue that "the photographer does not *make* his picture—A MACHINE DOES IT ALL FOR HIM," noting further that, apart from selecting the view, "*that machine drew* the picture for you."[32] This differs little from sentiments expressed by others who insisted, for example, that "the ordinary photograph is rarely, if ever, a picture. It might be an admirable photograph by accident, but that is no satisfaction to the photographer when his own conscience tells him that it was not he but the camera that made it."[33] Again, a fear that the whole process was dehumanizing haunts the cinematic apparatus by locating pictorial agency in the device and, more generally, in the process of mechanical inscription, and wresting it from the artist.

Like the cinema, once again, the phonograph (and to a lesser extent the photograph) have their stories of audiences literally fooled by a finished representation. Here, one need only recall the Victor Talking Machine Company's Nipper, whose belief in "His Master's Voice" long served as the very emblem of acoustic fidelity. Such stories seem less important to the present argument, however, than phenomena like the composite photograph whose combinations of several negatives could create a single seamless and unified world.[34] In photography, especially, it is rare to find stories where the image and the world are confounded as in the cinema, but the composite photograph's plausible world just as surely belongs to a representational paradigm distinct from those that stress object-causality or mechanical inscription as definitive. Like the diegetic world on the screen, which is convincing rather than confounding, the constructed spaces and times of a composite photograph (or combination print) like Henry Peach Robinson's "Fading Away" (1858) or Oscar G. Reglander's "The Two Ways of Life" (1857) distance themselves from the discourses associated with object-causality or with inscriptions. While compelling, their constructed worlds presume no *necessary* link to the actualities of either real objects or real acts of recording. The diegetic realm of filmic representation finds an even closer parallel in a more historically recent trend in phonography, where sonic landscapes constructed of dozens of discrete takes frequently bear no literal relation to any single, real performance, but are no less compelling for that fact.

Perhaps more importantly, while vernacular discourses of all sorts were segregating different causal circumstances of technological representation into different categories, they simultaneously pointed to the fact that one representational form might look like another (a "staged" event might look like a documented one). It was, therefore, often impossible to judge the referential status of any particular representation on the basis of its mere appearance. Given this fact, the significance of vernacular discourse is amplified, since by delineating causal circumstances these discourses implicitly defined the conditions under which, for example, a photograph might *legitimately* be taken as a document. "Diegetic" representations, for instance, threatened fictionality where they seemed to assert truth. Breaking without appearing to break the necessarily material, causal links between a representation and its referent made diegetic modes of realism (filmic or acoustic) convincing, ideologically powerful, and troublesome for representational cultures of all types. Delineating which were real and which were simply convincing was a major concern.

It seems plausible, then, that these stories, metaphors, and descriptions therefore serve primarily to define the semiotic potential of the technologically produced sign. In other words, while often couched in the languages of art and aesthetics, such debates had the force of analytic distinctions. These vernacular discussions served to identify, distinguish, and define a range of possible logical relationships between sign and object in the manner of C. S. Peirce's distinction between icon, index, and symbol, which produces a closely related taxonomy.[35] It is hardly accidental that these distinctions echo the trichotomy established by film theorist Étienne Souriau between the realms of the profilmic, the filmic, and the diegetic, as well.

The distinctions thus produced in public discussion are of great practical importance since, in addition to delineating regimes of belief and potentials for reference, they suggest specific forms of representational *practice* and particular modes of interpretation as well. If nothing else, they should lead us to complicate our overly restrictive notions of photographic and filmic representation along the lines suggested by André Gaudreault, who divides filmic narration into the distinct arenas of scenic monstration (mise-en-scène and staging), filmic monstration (framing and photographic processes), and filmographic narration (editing).[36] Thus, through a variegated discussion of image cause, the three realms of objects or performance, of inscription, and of diegesis emerged as conceptually distinct arenas for representational intervention—indeed, one might say that this discourse *produces* those realms as objects of knowledge. What is more, by producing these distinctions as a kind of informal, practical knowledge, vernacular and professional discourses about the new media effectively linked particular formal features like camera angle or blurriness to specific production practices and causal circumstances, like the conditions under which a negative was exposed. This knowledge thereby also more firmly linked the resulting representations to objects and events in the world. Simply put, the production and dissemination of this kind of knowledge effectively linked signifiers with objects and, historically speaking, produced the foundations for particular semiotic functions. Indeed, by regulating representational practice, this knowledge prepared the emergence of the "objective" representation.

Nor were these divisions limited to film or to entertainment. The points of narrational intervention Gaudreault derives from early cinema find an exact parallel in the representational practice of a Marey, for example, who characteristically enhanced his chronophotographs by working at each point successively. That is, to create a more legible image he invariably altered his object or mise-en-scène first, then refined his manner of inscription, and finally his means of resynthesis or presentation.[37] Marey was, furthermore, careful that none of his manipulations should affect the overall goal of representational objectivity, and thus restricted his interventions to those realms that would not affect its credibility. Phonographic practice proceeded in a like manner in the course of its own history.[38]

It is likewise essential to realize that strategies and techniques deemed appropriate to each arena were usually systematically interconnected. Again, Marey's case makes the point clearly. Only by *staging* events (in the realm of scenic monstration or the profilmic) in such a way that they might result in clearly *legible* representations could Marey legitimately lay claim to the "objective" properties of his inscriptional device. In other words, rather than manipulate the recordings or inscriptional process, which might have compromised the camera's objectivity, Marey manipulated the *object* so that he might otherwise observe the norms of experimental "nonintervention" and the demands of scientific objectivity.[39] As historian Marta Braun explains, when confronted with photographs of moving figures that were so detailed as to be illegible, Marey coordinated his techniques in the "performance" and "inscription" realms. "In the past, when Marey could not change an instrument any further to suit his subject, he would adapt the subject to suit the instrument," altering his performers' costumes so that they registered as "skeletons" on the negative.[40]

The emphasis that both informal and professional accounts place on the various processes of representational production (in contrast to reception) indicate the importance granted to causal chains of production for understanding the technical media's referential capacities. Hoaxers, photographers, and filmmakers had in different ways made clear, however, that mechanically produced signs could also lie, especially given the uneven distribution of technical knowledge among the general public. While widely disseminated and generally familiar norms of "straight" photography seemed to universalize and guarantee a particular kind of "documentary" reference, photographs that deviated from that norm parasitized its believability and dubiously promised the same referential security. Thus, because of its importance to representational epistemology, knowledge about image *cause* or conditions of production as it relates to representational practice and, by extension, to modes of referentiality, acquired the utmost social importance.

A photograph's ability to serve as reliable visual evidence, for example, depends on its indexical dimension, yet indexicality is not a simply evident "property" of every individual photograph. Instead, the indexical *function* is grounded in the knowledge of how some kinds of signs are typically *produced*, and therefore requires an inferential leap on the beholder's part. This inference assumes that the image has been produced under "normal" circumstances. Such inferences, while only probable, are, under force of habit, generalized to a larger class of formally similar signs. One need only produce an object (painting, digitally "doctored" photo, even a drawing) that *appears* to belong to the category "photograph" to falsely elicit the same inferential leap.

Public discussions of these media therefore served not only to distinguish the possible referential relations of, say, a photographic image, or a phonographic record, but simultaneously to distinguish the *conditions under which such relations legitimately obtained*. Given the increasing importance of photography to policing, jurisprudence, science, and news reporting—functions that no other image form could fulfill as effectively—it became increasingly necessary that particular logics of reference be clearly delineated and that *practices guaranteeing those logics* be identified and enforced. Stories about the new media's effects of immediacy were also stories of the conditions under which a technologically produced representation could be taken to function reliably as a *particular kind of sign* offering particular epistemological guarantees about its object.

If one understood, for example, the conditions under which a Marey chronophotograph had been made, one could trust its depiction of a particular process, even if it contradicted one's expectations. Consequently, in situations where such guarantees were of greatest importance, the relevant institutions developed precise strategies for regulating representational practice.[41] In essence, such regulations sought to guarantee that the representational practices an image *seemed* to imply had, in fact, been respected. The implicit fear that signs might not signify in a reliable manner suggests that particular formal features had, regardless of actual cause, come to be associated with specific production practices, and that these connections had solidified as *habits of interpretation or inference*.

Because the signifying relation between a sign and its referent is not verifiable by the sign's mere appearance, even "true" footprints or "normal" photos are irrevocably distanced and deferred from their referents. As supermarket tabloids continually remind us, a photo of Bigfoot (or Elvis, for that matter) or one of his footprints *can* be faked. In other words, the signifying marks by which we identify a particular depression in the earth as a "footprint," insofar as they are generalizable and can therefore authorize the reading of footprints *in general* as indexical signs, are conventional and iterable. Assumptions about sign production that sometimes allow us to track an animal through the forest or a muddy child through the house are the same assumptions that *could* be used to trick us. Likewise with a photograph. A photograph produced, developed, printed, and interpreted *under certain conditions* can, in fact, accurately document an object or occurrence, but another photograph, looking identical, but "faked," could very well "refer" in exactly the same manner, provided we apply the same assumptions. The reference of even a "normal" photograph is, therefore, always structurally (and *not* simply contingently) open to misfire. Moreover, this possibility is structurally necessary for even an honest photograph to refer in what we consider to be the "correct" or "normal" manner. Reversing the familiar hierarchy, the "normal" photographic relation is simply a special case of a more general referential possibility.

So, while inferential habits based on a *probable* referential status proved to be essential to the manner in which publics negotiated (and continue to negotiate) their day-to-day dealings with representational technologies, such habits were not untroubled. The various invocations of the commonplace "photographs do not lie," both affirmative and negative, that recur in newspapers and magazines throughout the last quarter of the nineteenth century, indicate a persistent skepticism regarding the technical media—a skepticism that could have profoundly material consequences.[42] For example, numerous legal battles were waged over the reliability of photographs, X rays, and films as evidence, in situations where it was necessary to eradicate epistemological uncertainty.[43] It is to the courtroom that I now wish to turn, in order to highlight once again how discourses about photography, phonography, and film insistently returned to questions of referentiality, and sought to delineate and discipline regimes of belief.

It is hardly difficult to find nineteenth-century melodramatic plots—theatrical or otherwise—that hinge on the revelation of an accidentally or surreptitiously exposed photograph in order to convict or exonerate a character of a capital crime. Boucicault's play, *The Octoroon* (1859), is perhaps the best-known example, where the unimpeachable evidence of a camera that had recorded the crime by chance condemns the murderer. Less well known, but present nonetheless, are similar plots that deploy the phonograph or motion pictures in the same role.[44] In contrast to *The Octoroon*'s simple presentation of the photo (" 'Tis true! the apparatus can't lie."),[45] the evidently truthful character of later apparatuses is not taken for granted. In a gesture reminiscent of Boucicault's, George William Hill has his 1883 *The Phonograph Witness*'s murder recorded by a hidden phonograph, installed experimentally to document all the business dealings of a bank president.[46] So dramatically troublesome are the epistemological issues at stake, however, that Hill spends several pages of dialogue explaining in elaborate technical detail how the phonograph works.

During the crucial courtroom scene, a long-missing phonograph is recovered and allowed to present its evidence. After having it explained that the device recorded the entire day's business, the judge asks if unimportant moments can be skipped, to which its inventor replies, "Certainly, Judge. The cylinder can be made, by pressure on this little lever at the side [*laying his hand on it*], to revolve with much greater speed." The discussion continues in even greater technical detail. The judge then asks how it can be returned to normal speed:

INVENTOR: Very simply. By raising the upper portion at the back, I can readily see the indentations on the metal sheet that indicate the sounds which occurred within the radius of the reproducing capacity of the Phonograph before they arrive at the point where they are reproduced. I can slacken the speed to its normal rate by simply raising my hand and relieving the lever of the pressure. (George William Hill, *The Phonograph Witness* [1883], 57)

Other equally precise indications are given in stage directions as to the technological specifics of the device, including the importance of a "clicking sound" while the inventor depresses the lever. In the end, of course, the phonograph testifies truthfully, the falsely accused man is vindicated, and the true killer unmasked.

Given the pot-boiling nature of the play, it is surprising that so much time is devoted to narratively unproductive and strikingly elaborate technical description, especially since *The Octoroon* foregoes such techniques in its invocation of photography. Perhaps Boucicault simply could assume a greater audience familiarity with the camera.[47] However, regardless of the actual state of knowledge regarding the phonograph, Hill's attention to spectator pedagogy seems remarkable and intense. In this play, a precisely detailed narrative of technological process serves to guarantee the causal conditions that render the phonogram a reliable record. It employs testimony in the role occupied by scientific ethics in the laboratory, and indicates with great force the more general role played by (informal) pedagogical discourses with regard to new media and their new referential possibilities. As in so many other instances, this discursive mechanism anthropomorphizes or, in a manner of speaking, rehumanizes the process of mechanical inscription by locating control over the device with an individual.

In fact, the courts, and jurisprudence more broadly, became a key site for analyzing the fallibility of the photograph, and for delineating the conditions under which an image might count as truthful or as fictional. Writers quickly came to realize that, despite its apparently ontological "properties" of truthfulness, the photograph was every bit the representation that a painting was. As one writer noted in 1897, "It is no exaggeration to say that an artist and practiced manipulator combined can do with the pencil of light pretty much the same as a painter who works with his brush and badger softener. . . . A photograph is not *necessarily* a faithful portrait."[48] Contradicting the rhetoric of autography, the rhetoric of art and artist clarified the extent to which the photograph was a "handmade," and therefore dubious, witness. Other articles and cases declared that the posing or arranging of people or objects, the framing of the image (which could eradicate undesirable elements), and the processes of development and printing could all undermine the photograph's apparent claim to truth. Indeed, the courts define the roughly same arenas of manipulation that we find in plays and anecdotes.[49] Indeed, its *apparent* truthfulness made it particularly troubling to the courts, leading the *Photographic News* to call photography "a most dangerous perjurer."

"Can the sun lie?" is often asked . . . a question which is supposed to carry its own answer. Perhaps we may say that though the sun does not lie, the liar may use the sun as a tool. . . . Let all then beware of the liar who lies in the name of truth.[50]

A 1909 courtroom story entitled "A Strange Witness" might serve to broaden the implications of this point, while simultaneously condensing many of this chapter's main issues and indicating its centrality to questions of cinematic persuasiveness. In this story, a photographer, observing a murder trial with a friend, insists he can determine its outcome. To the surprise of all, the next day he produces a film depicting the very murder in question. Explaining its provenance, he claims that in seeking to capture an "authentic" fight, he recorded his quarreling neighbors through a hole in the wall.

> JUDGE: Do you swear that the pictures to be shown here today were taken by yourself? [filmic]
> PHOTOGRAPHER: I do.
> JUDGE: Do they represent accurately the occurrences just as they took place before the machine? [profilmic/filmic]
> PHOTOGRAPHER: Sure.
> JUDGE: You swear to that?
> DEFENSE: Stop! . . . This is all irregular. *This machine*—I mean *this man* must be sworn![51] [causal confusion]

In the face of this seemingly unimpeachable evidence, a conviction predictably ensues, at which point the cameraman explains to his friend that he had, in fact, *staged* the film, based on details given in the previous day's oral testimony. In much the way that Hollywood has proceeded for the last seventy-five or eighty years, he simply worked from a script, included the signifiers of untampered recording, and produced the impression of a self-contained, real world that has been seen but not staged.

The premise here—that a feigned image could possess the affective immediacy of a genuinely "captured" one—offers a way of approaching Hollywood's strategies of diegetic immediacy. Diegetic, narrative cinema, in effect, adopts the formal features associated with specific practices of documentary or "actuality" production—practices of "capturing." In this manner, narrative cinema produces the effect of an independently existing world, oblivious to the fact of its being watched. If we extrapolate the context implied by the courtroom and apply it to the culture more generally, we begin to see clearly the stakes implicit in the seemingly innocuous stories of cinematic, photographic, and phonographic duping.

Mimicking the results obtained in this courtroom drama, the actual practice of filmmaking works its way from a notion of pure inscription to a more complex, diegetic understanding of film's potential. The idea that an event is *simply* recorded by a wholly external device, while a false cinematic ontology, is a key element of the ideological effect of technological representation more generally. As an article published in 1907 suggests, a crude documentary ontology seems to inform even those films that are staged. The motion picture camera, the author asserts, despite being capable of deceiving us with the illusion of *motion*, records only events that "have actually happened." "It is the manufacture of these *occurrences* so that the camera can *reproduce them* that is the most serious part of the motion picture business."[52]

Despite months and even years of experience to the contrary, producers routinely described their own practice as a matter of recording rather than representing. As a matter of discourse, the paradigmatic case of filmic representation remained the "captured" event. In response to a reader's inquiry into the use of motion pictures for recording news, for instance, a *Nickelodeon* editor pointed out that cameras necessarily arrived too late to be of real use because the world's events happen unpredictably.[53] In the same issue, another writer echoes this sentiment when he explains that outdoor scenes, for example, "are not, as is generally supposed, 'caught.' There is much planning and anticipation which goes into the 'catching' of an event."[54] The idea of "catching" an event somehow managed to condense or highlight attributes producers found to be characteristic in all films despite the fact that, since 1903, story films outnumbered actualities.

If compelling representation was understood to produce the effect of the "captured" event, how might those in the movie business have understood the role of staged events in the process? One writer described the situation this way:

> Formerly the manufacturer of motion pictures was obliged to content himself with an actual event such as a train in motion. . . . At the present day, however, it is entirely different. The work of the moving picture maker falls into two classes—actual events and acted *imitations of events.*[55]

It is clear that, conceptually, the role assigned to the camera remains unchanged here—it is still more or less "documentary," but now there are rel-

ative degrees of an *event's* authenticity. We can draw a few important inferences from this state of affairs. Producers assumed that the camera's recording of an event was in no way constructive of it—it simply recorded happen-
ings without a particular narrational point of view.

Gradually, however, alternative modes of representation began to emerge
and even the notion of inscription began to change. Commenting on a particularly rugged landscape "faked" in Brooklyn, one author reports:

> If the camera had swerved so as to take in a few feet to the right, the
> Western scene would have included a row of commonplace city flats. This
> is part of the cameraman's duty, not to swerve, to take in just so much of
> a scene and no more, or the whole effect would be ruined.[56]

This account clearly assumes that screen effects are not related to profilmic
actualities in any necessary way, and indeed, the camera may literally *produce*
a geography whose realities exist nowhere but the screen. The author recognizes that "realism" is not an absolute category when he asserts that "by
focusing his camera at just the right angle, the cameraman knew he could
make the scene *as realistic as he needed* for his purpose."[57] Rather than affirm-
ing that belief in the *film world* is ultimately at stake, however, the anecdote
ends when passersby are fooled into believing that a dummy is "actually"
drowning. In other words, despite recognizing the extent to which framing
produces a world that is "as realistic as it needs to be," and therefore need not
absolutely respect the integrity of an event independent of the camera, anecdotal confirmation of the film's realism comes from the domain of the profilmic—exactly the arena previously described as negligible in creating realistic effects.

In an intellectual gesture typical of later film practitioners, the requirements of an emergent mode of representation are here effectively translated
into the norms of an existing one. Rather than recognize that creating a fictional world may entail a credibility and coherence whose parameters are
defined within a purely filmic realm, these authors (and the many more who
tell similar tales) accommodate an emerging *diegetic* realism to a model of
cinematic representation based ultimately on a static *recording* of "real
events" which, theoretically, exist independent of the act of recording.

In discourse *about* the cinema, spectators are likened to passersby or witnesses and, correspondingly, the instrumentality of the camera is effaced. In
filmmaking practice, however, the constructive power of the camera is
repeatedly affirmed. The "rugged landscapes," "western scenes," and small

excavations made to appear as "deep gullys" that are shot in Prospect Park or Manhattan would never be believable were it not for their inscription on film.[58] The loss of an absolute sense of scale, which would be available to any profilmic witness, clearly belongs to a different logic of spatial and temporal articulation, which takes as its starting point, or point of coherence, the *filmic* (rather than profilmic) spectator. Crucially, the spatiotemporal logic necessary for achieving these types of effects requires a spectator whose spatial and temporal coordinates or points of reference come entirely from the film. In effect, the idea of cinematic representation as an act of pure inscription was confronted with and ultimately replaced by a more complex sense of the role of interlocking practices and an even more nuanced understanding of the relations possible between image and represented world. In short, an ideology of cinematic *recording* was acknowledging the fact of cinematic *representation*.

In the realm of phonography, the situation was structurally similar but charcteristically distinct. Still, given the general understanding of technological representation, those who were interested were encouraged to consider phonography as an act of inscription pure and simple. After the initial flurry of commentary on the phonograph's emergence had subsided (replaced in part by more topical discussions of motion pictures), the meaning of the phonograph seemed a far less urgent issue. Public discussion of the phonograph after about 1895 was comparatively calm, and whatever perceptual or epistemological dilemmas had troubled writers only a few years earlier appeared resolved, or simply forgotten in the face of the device's increasing familiarity. A vernacular form of aesthetic and epistemological "theory" nevertheless still served as the chief venue through which new understandings of the production and consumption of technological representations could be articulated, tested, challenged, and disseminated, but these debates diminished in both number and intensity. Despite the often-noted difficulty entailed in operating the early phonograph (which, despite his advertisements to the contrary, even troubled its inventor),[59] Edison's prediction that the *principle* of phonographic recording and reproduction would quickly become common intellectual property seems to have come true. In a fashion similar to the earlier dissemination of basic knowledge about photographic processes, writers for large city newspapers and middle-class, mass-market magazines typically assumed that the general public was familiar with the basics of phonography. As with photography, however, this familiarity encompassed a specific form of practical knowledge about phonographic production and a set of habitual assumptions about norms of making and using. Like photography, again, the phonograph was rendered familiar and predictable through a specific series of reifications. In short, in place of the camera's "click," which stood emblematically (and misleadingly) for the multiple processes of photographic representation, the moment of audio inscription, of "capturing," seemed fully to define phonography.

Producers and audiences seemed at least occasionally aware that phonography (as a representational process) encompassed the realms of sonic "staging" and reproduction as well as that of inscription. Its reification as an act of pure inscription nevertheless still characterized nearly all written discourse about phonography. Consequently, it enshrined the idea of simple "recording" as the defining relation of all phonographic practice—a feature of debates about sound with which we are still grappling. On this account, which is often assumed rather than stated, phonography transcribes sonic events that (although staged for the device) are fully autonomous of it. Notionally, these events would have occurred in exactly the same manner were the phonograph not present. In other words, phonography did not "penetrate" the event in any manner but sought instead merely to duplicate it from the outside. The only "representational" activity, therefore, was a kind of mechanical "listening," at once objective, sensitive, and passive. Given that most of the novel and perplexing aspects of phonography could be attributed to the agency of the device, the processes of staging the sound seemed, by contrast, less essentially "phonographic." In effect, defining phonography as inscription was a discursively efficient way to crystallize the basic epistemological questions raised by sound recording and reproduction.

While there was no great amount of explicit theorizing about sound recording and considerably less speculation about the phonograph's long-term impact between the mid-1890s and the early 1920s, several issues continued to dominate the practical development of the *craft* of sound recording. These issues thereby served as a kind of implicit theory of sound representation. While the explicit emphasis on perceptual simulation waned in discursive importance as recordists turned toward the more immediate technical necessities of obtaining sufficient volume and intelligibility, a general concern for perceptual fidelity continued to serve as a long-term guide to technical research and development. Indeed, writers projected a future in which no difference could be detected between a live musical performance and its recorded analog.

Taking this logic to its extreme was Edison, who in 1915 instituted a series of public demonstrations called "Tone Tests."[60] Performers such as Anna Case, who were under Edison contract, would tour with a phonograph and a set of recordings, perform beside them on a darkened stage, and challenge the audience to discriminate between the human and mechanical voices. These public demonstrations were apparently the very successful culmination of about a decade of similar, private tests performed by a variety of companies for their own staff or for critics and musicians.[61] As improbable as this may seem to contemporary ears, Edison clearly believed that absolutely faithful duplication had become the obvious and paradigmatic representational goal, and he enshrined it as his chief marketing strategy. In fact, Edison insisted that his "Diamond Disc" series be sold not as records but as "Re-creations." The Victor Talking Machine Company, too, had since 1902 based their most successful marketing campaign on their ability to duplicate Caruso's voice. Typical Victor advertisements showed paired photos of tenor and disc and boasted "Both are Caruso." One such ad continued,

> The Victor Record of Caruso's voice is just as truly Caruso as Caruso himself. It actually is Caruso—his magnificent voice, with all the wonderful power and beauty of tone that make him the greatest of all tenors . . . When you hear Caruso on the Victrola in your own home, you hear him just as truly as if you were listening to him in the Metropolitan Opera House.[62]

Regardless of whether any recording actually achieved the desired ideal or ever could, the currency of such an explicit standard helped to shape the popular and professional understanding of phonography's raison d'être.

Nevertheless, Edison's recording *practices* suggest a very different understanding of phonographic representation. However much he may have promoted the slogan "Comparison With the Living Artist Reveals No Difference,"[63] he clearly assumed that live performance and phonography were different enterprises altogether, involving techniques and standards appropriate to each medium. Nowhere was this distinction more evident than in Edison's peculiarly undistinguished roster of singers. For years, critics and former employees have marveled at and usually criticized Edison's idiosyncratic taste in choosing musical talent for the phonograph. Compared with the stellar lineups of celebrity opera singers at other companies, Edison's stable of performers was singularly lacking in name recognition and critical esteem.[64] When pressed by business partners and employees to contract better-known performers for their own series of vocal cylinders/discs, Edison famously rejected nearly every singer of stature in the world. He became notorious among producers and collectors for adamantly refusing the most highly regarded in favor of relative unknowns and nondescript journeymen. Moreover, he reveled in his decisions to reject the highly touted and jealously maintained the purity and superiority of his own requirements.

When queried, the nearly deaf Edison insisted that all his phonographic talent satisfy his own personal auditory standards, which were attuned not to musical norms but, he argued, to the demands of the machine:

> I am like a phonograph. My ears, being a little deaf, seem to catch all the use-less noises more readily than the musical tones, just as the phonograph exaggerates all the faults of a singer. . . . From a mechanical point of view, music is in pretty bad shape. . . . There are too many interpreters of music and too few persons interested in perfecting its mechanical deficiencies. . . . My deafness has aided me in my work not only because it has the effect of giving discordant sounds an exaggerated prominence in my ears, but also because it has taught me not to rely upon my own hearing for judgment. Consequently, I have made recording instruments more delicate than human ears showing beyond question whether a singer sings in tune, whether a voice has tremolo or whether any false note or sounds are present.[65]

Musically untrained, Edison's faith that he could divine false notes or designate inappropriate vocal effects is surprising but understandable. His sense of aesthetics was grounded less in music than it was in mechanics. Phonograph historians John Harvith and Susan Edwards Harvith note that Edison "considered dramatic personality intrusive on discs and developed a stringent, mechanical perfection aesthetic for recordings that included purity of tone, extreme clarity of enunciation, and the abolition of extraneous noises, which, he conceded, would not be objectionable in the concert hall or opera house."[66] Edison detested tremolo most of all, preferring absolutely unornamented voices, and ones that were "attuned" not to large concert halls but specifically to the medium of recorded sound. His sense of artistry was determined by the quality of the phonogram and not at all by the performance that produced it.

When finally convinced to hire top-notch singers for his Diamond Disc series, Edison made himself a reputation as a difficult, tin-eared, and harsh taskmaster. He infamously criticized Dame Maggie Treyte (as he had Caruso), for the "quality of her top notes." At first, she understandably regarded his opinion as ignorant or uninformed, but later came to take his insistence on vibrato-less singing as an attempt to get her to " 'adjust' to the recording characteristics of the day," and thereby register with greater clarity on the disc.[67] In a similar vein, an Edison staff pianist remembered that the inventor also refused to record music with complex harmonic structures, preferring simple thirds and sixths over extended harmonies, which tended to muddy the resulting record.[68] Indeed, even Edison rival Fred Gaisberg, who first recorded Caruso for the Berliner Gramophone Company and subsequently made him the most famous singer in the world, recognized that not all singers could record equally well. He went so far as to credit Caruso's extraordinary success to his performance technique and singing style, which together produced "the one perfect voice for recording."[69]

As testimony from Anna Case (later to star in one of Warner Brothers' earliest Vitaphone sound films) and other singers makes clear, under Edison's relentless pressure they came to understand that sound recording was a process that could involve more than passively duplicating what was effectively a concert hall performance. Performers and technicians alike learned that it might even be advantageous to change aspects of a particular musician's performance style in order to take advantage of the machine's peculiarities. Through repeated exposure to the most mundane and practical aspects of musical recording, both groups came to believe that sonic representation was not simply a matter of precisely transcribing completely prior and autonomous events, and to concede that a performance, for instance, might deviate from its customary *presentational* norms in order to achieve a particular *representational* effect, like intelligibility, regardless of its effect on the character of the "pro-phonographic" event.[70]

This was no minor realization, but one as transformative of phonography as the process of diegetic, continuity editing was for the cinema. In effect, Edison and his cohorts discovered that the specific acoustic qualities of the pro-phonographic performance were fundamentally irrelevant to the phonographic process, as long as the final cylinder or disc "reproduced" (in fact, "produced" would be a more accurate term) a satisfactory performance.[71] Gradually they came to realize as well that, beyond performance, the processes of inscription could similarly *create* sonic effects, and even reproduction or playback could enter into the complex calculus of sound representation. By emphasizing the sonic character of the final product over all other considerations, Edison gradually weaned his performers and technical staff from the prejudice that inscription was the only properly phonographic act, instilling in its place a flexible and essentially "diegetic" understanding of sound representation that created coherent but spatiotemporally nonliteral musical worlds.

Nor was Edison the only producer to leave his mark on the emerging theory and practice of sound recording. In fact, several other important figures were arguably even more influential in shaping the way our musical world sounds. In particular, the field of classical vocal recording deserves special consideration, since it was here that musical and technical standards were in practical terms most demanding and most explictly debated. It is likewise the classical recording that was most frequently at issue in the proto-theoretical discussions that occurred in popular magazines, technical reports, and journals like *The Gramophone*.

Fred Gaisberg, an accompanist and record producer at the Berliner Gramophone Company, the Gramophone Company-HMV, and later EMI, set the standard for the earliest style of classical music recording. In one writer's words, "His conception of his role was not grandiose. He was an engineer and a businessman, charged with getting the best musicians to record and seeing to it that the disks were without serious blemish."[72] His successor, Walter Legge, subsequently redefined phonography as a collaboration between musician and producer that sought to produce a performance at a level of perfection unattainable on stage. Legge was notorious for shaping and fine-tuning performances to a degree unimaginable in regular performance, through minute rehearsal, and for using the accumulated library of previous recordings as a source of ideas to be mined, culling from them the techniques that worked best on disc. Finally, he exercised more control over the selection of recorded takes for release, trusting his own studio-sharpened ears over those of the artist alone.

Later, John Culshaw was to update these techniques by severing the finished representation from any real or realizable "original" performance, no where more dramatically than in his 1958 recording of *Das Rheingold*, which convinced thousands of listeners of the importance and aesthetic possibilities of stereo production.[73] Emphatically insisting that records are listened to in the home, and not in a concert hall, he created a sonic world more "perfect" or more fully realized than could exist on stage. While often recognized for his attempts, in the *Rheingold* recording, to re-create the dramatic use of stage movement in live performance, he also went far beyond the possibilities of the stage. He suddenly and dramatically grants particular characters an impossible and superhuman spatial ubiquity, and actually uses the nine anvils and seven harps that Wagner suggested, but which no company could reasonably accommodate. Finally, he incorporates the thunder and myriad other sound effects demanded by the text.[74] Culshaw's 1968 *Elektra* was even more daring (and more controversial) because it relinquished the theater completely, powerfully reimagining operatic space and sound in terms appropriate only to the record, and wholly impossible on the stage.[75]

The most thoroughly and self-consciously "diegetic" understanding of sound representation, however, emerged in the late 1930s, unsurprisingly in contact with the cinema. The collaboration between Prokofiev and Eisenstein on the latter's *Alexander Nevsky* (1938) prompted the composer to reimagine musical recording along the model of Eisenstein's visual techniques.[76] Prokofiev argued for a recording strategy that combined different sonic "angles" and "scales" in order to create a whole that was as fully penetrated by the microphone in its conception as Eisenstein's visible world was by the camera. For the most part, Prokofiev's "narration" of the music consisted of "unnaturally" foregrounding particular instruments by placing micropohones very close to them, or recording different sections of the orchestra in different studios, producing effects the composer likened to standing the orchestra on its head.[77] During recording, Prokofiev stood in the mixing booth rather than on the podium, overseeing the performance through the mediation of the recording apparatus rather than the mediation of the concert hall. Indeed, for the composer, the piece was as inconceivable without recording as Eisenstein's films were without the camera. Neither considered representation a process of bringing an external device to bear upon an otherwise complete performance, but instead imagined staging, performance, inscription, and editing as inseparably united aspects of the same integrated vision.

So radical a rethinking was this (although so unobtrusive sounding) that even the ostensibly more radical sound theories of the Soviet filmmakers give it little attention. Nor did it have much of an effect in more purely theoretical realms. Given this long tradition of practice to the contrary, it is somewhat surprising how the academic literature about sound in film is so thoroughly permeated by the assumption that sonic representation amounts to nothing more than absolutely duplicating some event that is otherwise fully independent of the act of mechanical inscription. The reduction of a complex system of representational practices to a single act of mechanical recording or inscription is characteristic of nearly all representational technologies, and derives from an understanding of representation as an act of mechanical perception. The mechanical nature of the devices, and the uncannily precise effects of nonhuman technological recording, so thoroughly crystallized the novelty and epistemological uncertainties embodied by the new media that it is hardly surprising that the camera's "click" and the microphone's passive "listening" came to stand for the totality of the process.

Even if the preternaturally present images and sounds of the new media suggested a world where machines could perceive and preserve a world that produces its own record, the varied practices of image and sound representation within particular institutional settings and under specific representational demands proved again and again that even the single image or recording implies a broad range of different possible stories of production. The most convincing "Arizona" landscape might be shot in Brooklyn and require framing out Coney Island attractions, and the most "faithful" recording of a Caruso might require him at times to sing like an amateur. As Marey proved in one way and Edison in another, technological representation is never a case of simply *seeing* or *hearing*, but of *looking* and *listening*. We look and listen *for* things, for specific purposes, while the machine's "more perfect" eyes and ears simply absorb indiscriminately. Whether recording muscle contractions for a study of human physiology, or an aria for the most demanding audience, it is not possible to separate the performance from the processes of inscription, nor neither from the conditions of exhibition. Even the most "objective" recording cannot leave its object untouched, for the very goal of objectivity touches the core of that which we wish to record, and the techniques we use to record it.

Thus the three realms I identify as characteristic of all forms of technological representation (whether or not we call them performance, inscription, and diegesis) must be understood together, even in those instances where one seems utterly unmanipulated. In the chapters to follow, I hope to bring the historical insights of the first two chapters to bear upon a very specific historical period, representational form, and institutional setting—the "coming of sound" to the Hollywood cinema—concentrating on the interaction of institution, representational practice, discourse, and devices. In spite of the lessons of the nineteenth century's various encounters with the photograph, the phonograph, and the cinema, the actors in the drama of Hollywood sound themselves stage and restage its debates, discover and rediscover its insights. As in the first two chapters, I will limit myself to the "inaugural" era of a new technological form, although here the form is less defined by a wholly new device or technique than by the intersection of two technologies in a new, multisensory hybrid, the sound film.

CHAPTER 3

EVERYTHING BUT THE KITCHEN SYNC
Sound and Image Before the Talkies

It is the purpose of the remaining chapters to examine the factors conditioning the emergence of "sound" within the American cinema. I will demonstrate not only how these events recapitulated several of the key features I associate with the emergence of the technical media more generally, but also how entirely new and unimagined possibilities may emerge out of a limited and concrete set of historical and material conditions. More specifically, in this chapter I examine how the entities "the cinema" and "sound" came to be defined in relation to one another historically, and how the very devices and institutions we associate with the Hollywood cinema were similarly defined and redefined over time. While the "coming of sound" may seem a particularly clear-cut instance of adapting a well-defined technology to a similarly stable aesthetic form, such is decidedly not the case. Not only were both technology and representational form in flux, but each helped to define or constitute the other in the process of their mutual interaction.

Moreover, the period before the putative coming of sound offers us a glimpse into how two new technologies with sometimes overlapping and sometimes quite distinct histories—namely, cinematography and phonography—could combine to form an integrated sensory experience that was neither audio nor visual, but distinctly audiovisual. However, the proper ratio between the senses—between hearing and seeing—was open to vigorous debate and competing models. Was the cinema an essentially aural medium, born of the spirit of publicly performed music, to which spectacles of various sorts might be appended, or was it a narrative visual form, to which sound could only be an "accompaniment"? During the early cinema period, especially, we can indirectly witness a confrontation between competing sensory regimes, each one adhering to its own dictates and coexisting at times unpeacefully. The dominant ideal of listening as attentive, sensitive, and receptively passive—as complete and self-contained—confronted the film industry's emerging visual language, and each sense was forced to work with and against the other in a project of audiovisual collaboration. As the institution of cinema moved more and more firmly toward narrative, sound took its place as an integral, but also overdetermined, part of the sensory order characteristic of the classical cinema.

SOUND IN 1926

In retrospect, the sonic needs of Hollywood in 1926 seem almost obvious. Who, for example, could ever have doubted the primacy of intelligible, narratively important speech for the classical paradigm? The cinema of narrative integration, long since established, apparently mandated the practices of sound we have come to accept as the norm, as it simultaneously ruled out other approaches. David Bordwell, Janet Staiger, and Kristin Thompson, for instance, assert that "sound as sound . . . was inserted into the already-constituted system of the Classical Hollywood style," that "the centrality of speech became a guide for innovation in sound recording," and that the use of character-centered leitmotivs indicates how even musical accompaniment had become a fully integrated component of a narrative unity.[1] Recorded sound simply fulfilled already existing functions, and indeed was their "functional equivalent." Change, then, was almost entirely technological.

However, the situation appears this tidy *only* in retrospect. It seems probable that the leitmotiv system was not universally adopted, or even dominant, nor was the function of sound accompaniment of a single type or exclusively narrative in intent or effect. Indeed, the very term *sound accompaniment*, which assumes the primacy of the image in all cases, obscures our view of the so-called pre-sound era and, furthermore, orients us toward the classical cinema in such a way that we may misrecognize alternative traditions simply as fully integrated components of the classical style. Indeed, the perspective that does not expect to see discontinuities or complexities may not perceive particular deviations as deviations at all, encouraging us to notice those elements that led most directly to the ultimately triumphant set of norms. The archive and the genealogy that I have produced in the first two chapters and in the present one encourage a different focus, and different conclusions.

As recent research confirms, the few years immediately preceding the "dawn of sound" are too narrow a historical slice by which to characterize phenomena as varied and complex as "film sound" and cannot adequately characterize the evaluative and practical frameworks within which sound was received. Although terms such as "cinema sound" appear to name objectively describable entities defined by the technologies that "produce" them, they are, in fact, not neutral descriptions.[2] That is, the act of naming or describing determines the boundaries of the phenomena under discussion and effectively creates the contours of its object, and basic terms like *synchronization*, *music*, and *effects* may not always have meant what they have subsequently come to mean. The boundaries between media, and even between technologies, were and are open to dispute and renegotiation, as are the boundaries of *function*.[3]

The title of this chapter evokes the heterogeneity of practices, functions, and effects of sound in the cinema, and hopes to open up calcified and often unexamined uses of the term "synchronization." To start, let me suggest that any fixed or purposeful relationship between sound and image legitimately may be thought of as synchronized. From this perspective, "lip-synch" is only one, rather banal, possibility. Consequently, I will avoid approaches to synchronization that focus exclusively on *production*, and thus ultimately on the relationship obtaining between sounds and sources in the everyday world, in order to ask if the represented sound and image correspond to the sound and image relationship obtaining in the "real world." Theories of sound representation that conflate the profilmic and diegetic realms by reducing representation to inscription reinforce this attention, as do approaches that base their notion of synchronization on the naturalistic norms of later classical cinema. Our current view of synchronization and, more broadly, our current theories of film sound too often derive entirely from a consideration of sounds linked to pictured sources—their presumed real-world sources—as if representation were merely a matter of duplicating that world. As chapter 2 implies, however, it is better to focus on representational *effects*, putting the emphasis back on the film experience and moving it away from potentially irrelevant issues concerning, among other things, the arena of the profilmic. By expanding our sense of what it meant for sound and image to "go together," I hope to suggest how the historically shifting roles of sound have helped to define and redefine the mode of representation, the mode of production, and the pleasure characteristic of the Hollywood cinema at various historical moments.

In their seminal analysis of the classical Hollywood cinema, Bordwell, Staiger, and Thompson forcefully elaborate an account of the classical stylistic paradigm as a structured and structuring framework regulating the boundaries of aesthetic options and decisions. Arguing for a more or less "functionalist" approach to the analysis of textual production and systematicity, they discuss formal devices in terms of their specific roles within a historically and culturally defined set of textual possibilities.[4] Thus, they argue, given the historically limited means (both technical and imaginative) available at any particular time, it is possible to reconstruct, through careful analysis and documentation, which formal devices could at any particular moment, and in any particular textual situation, substitute meaningfully for one another, the way a straight cut can substitute for a dissolve or a fade-out and fade-in. From this perspective it is easy to understand how and why certain textual forms persist despite possible alternatives, as well as how aesthetic options are ordered hierarchically in terms of probable use.

While powerful as an analytical tool, the functional equivalent, as a *practical* concept, is central to the organization of the Hollywood industry and its mode of textual production, which takes the continuity script as the basis for the realization of the cinematic text.[5] This approach affirms the narrative and even verbal origin of all cinematic representation and structures the type and frequency of formal devices employed in the text. Unlike the earliest cinema exhibitions, which often reveled in the absolute specificity of each and every take, even encouraging multiple viewings in a single session,[6] the mature cinema of narrative integration depends on the assumption that five filmed versions of the same narratively mandated event are virtually interchangeable. Since each shot or device is determined by the narrational function it is to fulfill, the field of representational possibilities is hierarchized in advance, and the possibility that two realizations of that function greatly differ is dramatically reduced. What the individual shot loses in phenomenal specificity, however, it gains in narrative saliency. The more regulated the relationship between script, realization, and final text, the more interchangeable the personnel can be as well—an economic benefit that cannot be overestimated.

The emergence of the classical cinema as a form of practice therefore required nothing short of a radical redefinition of the cinematic image. From an essentially idiosyncratic form that relied on a "topographic" mode of spectatorship, which scanned the entire image, and a highly presentational mode of exhibition that emphasized performance, there emerged an essentially narrative pictorial form whose textual specificity was removed from the hands of the exhibitor and placed, as much as possible, within the mechanisms of the text itself. Despite this trend, however, until the sound era certain aspects of the film experience—like sound and music performance—remained beyond the producers' control.[7]

Like a modern printed text, whose typographic idiosyncrasies are of almost no importance in comparison with its signification, the filmic image became a relatively transparent signifier of a verbalizable event, rather than an "event" in and of itself.[8] Whatever it thereby lost in complexity and "grain," it compensated with a gain in legibility and signifying stability. Particular filmic devices such as dissolves, fades, and wipes similarly abandoned whatever phenomenal appeal they might have had in favor of their ability to signify certain kinds of spatial and temporal articulations in a consistent manner. Within such a textual paradigm it seems reasonable to assume that, at least at a practical level, film *producers* evaluated devices in terms of their functional equivalence, and more or less ignored those aspects that were clearly nonequivalent.

When examining the state of film language at any specific historical moment, we are able, through the concept of functional equivalence, to analyze devices in terms of their paradigmatic substitutability, but it is less successful accounting for difference and contradiction. As a tool for explaining historical change, it can even be misleading. Bordwell, Staiger, and Thompson argue that Hollywood successfully innovates and adopts new representational technologies when they fulfill existing textual functions. This suggests a fairly uneventful sequence of problem and solution, with certain technologies succeeding because of their evident suitability to the classical narrative, others failing because they offered no functional gain. Certainly there is a great deal of truth in this position, especially given Hollywood's general conservatism in textual form and in investments in new machinery. All the major technological innovations, like sound, color, and widescreen processes, did indeed mesh with existing textual forms and functions. However, to think of these innovations as *equivalently* fulfilling certain functions robs our historical explanation of an important degree of complexity.

To the extent that, say, recorded sound did not precisely fulfill an existing function, it introduced an element of excess or contradiction that cannot simply be ignored or discounted. In fact, such contradiction seems generative of textual and even industrial transformation. To complicate matters further, given the multiplicity of possible textual functions for sound, to suggest that sound fell into a preordained slot, or even group of slots, is unnecessarily reductive.[9] Even while appearing to fulfill a particular function (such as providing intelligible, narratively important dialogue), sound may simultaneously be performing other, nonnarrative or even nonrepresentational ones. The standards of sound representation eventually codified by Hollywood practice, although appearing to be yet another familiar set of techniques that aids the effortless assimilation of story information, and therefore simply another component of the classical narrative system, conceal a history of traditions of representation and exhibition that may be at odds with classical norms, and at times are in direct conflict with them.

Sound technicians, and what might be called "sound practitioners," will be my direct focus in this and subsequent chapters since it is through them, their goals and standards and their systems of practices, that these varied traditions are realized. Rather than a simple matter of problem and solution, I see the innovation and adoption of new technologies as a process fraught with controversy and struggle—the stakes often involving the professional identity of the technicians involved.

In general, filmmaking followed a historical course that is by now familiar. After approximately 1907, narrative films dominated not only in sheer numbers but as the form around which the emerging industry grew. After that point as well, the heterogeneous and often boisterous field we call early cinema moved toward greater and greater formal and institutional unity. Along with the rise of narrative films came a more pervasive pressure toward narrative integration that encouraged producers and exhibitors to subordinate both filmic and extra-filmic elements to an overarching narrative effect, and thereby unify and regularize the cinema. Sound accompaniment was probably the element of the show most resistant to these pressures because its functions and effects were only occasionally narrative. In excess of their narrational functions, music, lecturing, and effects spoke directly to audiences in a celebration of their fleeting community. Whether through sing-alongs, virtuoso musical or vocal performance, ethnic or national music, conspiratorial mockery of the film, or simply their bodily presence, musicians and sound performers decisively mediated the experience of silent films in the nickelodeon and in the picture palaces. While few of us can recall the experience firsthand, it seems certain that sound and image coexisted in many more variations than in the talkie, and that sound might, at times, exceed its role as mere adjunct.

By briefly sketching the history of sound and image combination in the cinema, I will show how particular sound practices that apparently disappeared with the advent of the talkie continued to exist in the margins of the classical style. Many of these practices belong to a mode of cinema that might be characterized as more presentational than representational, and therefore as belonging to a vision of the cinema—a cinema of attractions—alien to that shared by the sound technicians of the transition period, who were more uniformly immersed in the narrative norms of the classical cinema.[10] In order to address these concerns, we need to interrogate our idea of accompaniment by asking, what was "sound," what was "synchronization," and what was understood as "a film"? As I will show, debates about these questions work to overcome conflicts between traditions of practice and institutional demands by bringing them into agreement, however tenuous that agreement might be.

SOUND'S DIRECT ADDRESS

That the silent cinema was never silent has become one of the great revisionist clichés of film studies. In point of fact, this is not quite true, for in many cases sound was an integral part of the nickelodeon program *except* at the time during which the film was projected.[11] At many nickelodoens, music was played during the intermissions and with the obligatory illustrated song,[12] but sound's most important early role was as a means of attracting attention to the nickelodeon itself. Sound, in the form of phonograph, barker, drummer, or piano, primarily served the function of hailing potential patrons. As nickelodeon historian Q. David Bowers notes,

> Apart from numerous attempts to use phonographs to provide soundtracks for movies, the primary use of the record player was to attract patrons into the theatre. Thousands of nickelodeons had phonographs in the projection booth, with the horns projecting out through a wall into the street, playing through an opening usually above the ticket booth. The music thus provided was loud enough to attract passersby directly in front of the theatre but not sufficiently loud that merchants would complain up and down the street.[13]

In 1910, *Film Index*'s music columnist Clyde Martin concurred, telling his readers that he was recently fired as a piano player "because they could not hear me on the street. . . . That is the fault of the average exhibitor today; he doesn't want a piano player, he wants a Bally-Hoo." Martin's rival at the *Moving Picture World*, Clarence Sinn, similarly recalls, "When music was first introduced to the picture theater, they 'wooped 'er up' until the music could be heard out on the street."[14] Phonographs themselves were often called "barkers," since they were so closely identified with this role.[15] The hailing function of sound became so common, in fact, that several cities enacted legislation prohibiting the use of mechanical sound devices outside nickelodeons,[16] and contemporary descriptions of theater design stress the importance of shutting out external noise, including barkers and the like.[17]

However much "barking" may have dominated early film sound, other sources from the same period suggest that in some theaters music accompanied *only* the film.[18] Whatever may have been the case, it is important to realize that cinema, as a cultural practice, was not dominated by a single approach. As Rick Altman argues,

> Silence was one of a number of acceptable film exhibition approaches throughout the pre-1910 era. Indeed, in certain regions and in smaller theaters, silence during the film remained the rule even after 1910; in other situations, and for a limited number of years, silence was simply one of many operative strategies. In either case, audiences were not generally trained to *expect* continuous musical accompaniment of films until after the turn of the decade." ("The Silence of the Silents," 677)

Until the early 'teens, sound worked to solicit patrons and draw their attention—a function that was to support a more general strategy for directing spectators to performers in the theater, or to objects in the image. In fact, the direct address made possible by sound would shape most silent film sound in one way or another.

SOUND OVER IMAGE: THE STRUCTURAL PRIORITY OF SOUND

Beyond "barking," which went on regardless of whether an image appeared on the screen or not, the best documented and most widespread audiovisual form on the nickelodeon program was the "illustrated song," which came to prominence in the entertainment world in the 1880s and 1890s. The business flourished in the 1890s with the advent of the amateur photography boom.[19] These songs, while a staple of nickelodeon programs, were not designed solely to entertain the patrons of the theater. Sheet music publishers routinely provided nickelodeon owners with free (but apparently inferior)[20] slides in order to get them to perform and thereby promote certain new songs.[21] While truly audiovisual, these were essentially *sound* performances that encouraged sing-alongs; the visual presentation was decidedly secondary. Song promotion and sheet music sales drove this practice and organized its formal properties, putting music ahead of images. In economic, conceptual, and structural terms, the song—*sound*—was clearly dominant over the image.

The illustrated song has an important and ultimately more influential parallel in the slide lecture. The slide lecture, often a travelogue, holds a privileged place in the history of cinema for several reasons. The collection of views typical of the travel lecture, which were united into a continuous whole by the mediation of the lecturer's account, and particularly by the fiction of a journey, offered an early and persuasive model for multishot narratives in the early cinema period. As Charles Musser has shown, the travel genre was one of the earliest proto-narrative forms of cinema because it offered a compelling logic that motivated and unified the succession of otherwise disconnected stills and films.[22] In addition, by proposing an embodied observer located *in* the fictional world, this form paved the way for the subjectivization of space in the Hollywood classical narrative through devices such as the point-of-view shot.[23] Significantly, travelogues were a sound-dominant form.

Subjectively organized spaces and identification with the traveler were implicit touchstones of many early accounts of travelogues (which began to include motion pictures as well as slides),[24] and in "how-to" articles in the trades. Lecturers were prized for their ability to create a sense of vivid immediacy. A tour of Yellowstone, for instance, merited comment because it was presented by a stagecoach driver intimately familiar with the area, while a series of talks on Washington, D.C., New York, and Montreal likewise found favor because the lecturer had gone to these locations himself to acquire the views. John Stoddard and Burton Holmes, acknowledged masters of the genre, were singled out for the quality of their images, which "harmonized" with their lectures because they designed and shot their own slides to match the narratives, rather than employing stock images.[25] So popular were these travelogues that, in 1908, the Keith theater chain was even using them as headliners.[26]

Lecturing can tell us something about the role of music, too. Like the slide lecturer, the singer provided a sonic, and often narrative, pretext and continuity for a series of images, either stills or films, which illustrated the song. As in the practice of lecturing, *images* here accompany *sound*, not vice versa. The functional subservience of image to sound is emphatically declared by the widespread practice of illustrating songs with stock slides rather than with specially produced series, indicating that the motivation for the performance was aural, not visual.[27] In fact, it was argued that the best lecturers never openly acknowledged the images but rather spoke continuously while an assistant changed the views at the appropriate time.[28] Images were not irrelevant to the process, however, and both singers and lecturers met criticism when their slides did not live up to the sounds they were meant to accompany. Successful lecturers like Stoddard, Holmes, and Dwight Elmendorf stood above the competition by upgrading the images that accompanied their talks. Unlike many of their competitors, all three traveled to the locations they described in order to take photos that would therefore more precisely match the spoken descriptions. This insistence on better "matching" was later applied to song slides as well, and standards of matching offer a glimpse into early notions of synchronization.[29]

Discussions of successful "matching" in lectures and illustrated songs stress a "harmony" between sound and image. Harmony, whether understood as illustrative accuracy or as thematic consistency, in fact describes another fairly stable tradition of synchronous relationships which has been submerged by our current limited perspective on sound and image possibilities. The praise associated with "harmony" was also accorded to other forms of synchronization, including ones that emphasized sound performance at the expense of the image. In lectures and songs, sound dominates and the image provides the momentary "shock" or "effect" that enhances the lecture. The lecture or the song would no doubt be most effective when sound and image "fit" or "synchronized" together, but as the use of stock slides and footage illustrates, it was the image that was fit or synchronized *to* the sound.

Beyond the lecture, promoting the sound component of the motion picture show seems to have been good business for exhibitors, even if manufacturers had already begun the drive toward narrative integration and image dominance. Writers from the 1907–1911 period repeatedly stress the differentiation, distinction, competitive edge, and more refined audiences available to those who added sounds, and especially music, to their programs.[30] The prominence of sound at a "picture" show was no accident. Indeed, the idea of sound-track dominance influenced the very structure of at least one form of filmmaking. Of the early forms of "talking picture," the ones that occasioned the most hope and, apparently, the most premature praise were the myriad phonograph and projector combinations. Going under the names Cameraphone, Chronophone, Theaterphone, Biographon, Cinephone Synchronizer, Phono-cinematograph, and Photophone, to name but a few,[31] these variously constructed and variously successful portmanteau technologies were extravagantly praised as mechanically and aesthetically perfect, eliciting predictions of the imminent demise of the silent picture. Unfortunately, none of them adequately solved the problem of synchronization, although the American Cameraphone and the French Chronophone did enjoy a certain degree of success.

One early example, the Phono-cinematograph, was described in 1907 as a "thoroughly practical success," although its synchronization system depended on the capabilities of a projectionist with a talent for on-the-spot adjustments as a part of his performance. Significantly, the device required the performers to record the sound first and *then* film a performance that matched the sound track, much like our current practices of lip-synching.[32] Other companies followed the same procedure, selecting a "stock" record from any record store and filming a series of images to "fit" it.[33] The most common practice was to record a musical performance or an excerpt from a play and then "drill" the performers to "match" the record before filming the picture's track.[34] In the noted case of an early "Motographic Opera," selections from *Il Trovatore* were recorded on a phonograph and images later matched to it. Advertising and critics' responses make it clear that the overall form of the production stressed the musical resumé above all else, rendering the image incidental. Like many other sound practices, this sort of "synchronization" carries a distinctly performative flavor, but more important, these disc and film combinations all treated the sound recording as the structural basis of the entire representational amalgam.

Discursive, participatory, emphatically present, and lucrative for exhibitors, sound constantly threatened to usurp the image. The priority of sound over image became a governing structural feature of other types of films and in other exhibition practices as well. Often, in advertisements, producers would stress a film's opportunity for "effects," that is, for clever musical or percussive effects to accompany the image. While this may frequently have been limited to "being unduly noisy when a man falls from his horse,"[35] in many cases the sound practices were more sophisticated. In an area of great theater concentration, like New York's Union Square, at a time when film production could not keep up with demand, the success of a theater often depended on qualitative difference in presentation. Nickelodeons would compete by providing the most sophisticated, elaborate, or funny sound performances to fit a particular film. Articles suggested that middle-class "girls" who played "high class" music could draw high-class patronage as well.[36] Others counseled that "props," or sound effects, were the best way to improve bad pictures,[37] while still others suggested that, measured in profit margin, sound effects were a sound investment.[38] These strategies placed an unusual emphasis on performance, discursivity, and "liveness," and both lecturers and "prop men" understood "synchronization" as a category of evaluation. Contemporary authors therefore repeatedly encouraged lecturers to "roadshow" their lectures to get both text and timing just right. The many articles praising Lyman Howe's presentations use performance criteria in just this way, and ascribe his success to his troupe's perfect performance.[39]

Companies even began to produce films whose purpose was to motivate particular forms of sound accompaniment. A typical example is Biograph's "Fights of Nations." The film depicts a series of different racial and ethnic types who come into conflict, but who are reconciled in the end. The use of recognizable ethnic types provides a pretext for clever sound synchronization. As each type appears, the accompanists play a melody evocative of his or her homeland. As types enter into conflict, the themes compete for dominance and collide in discord. As the advertising suggests, the music underscores the conflict's resolution into "harmony" by playing "America."[40] "Fights of Nations" is by no means an isolated instance. The trade press abounds with films whose chief purpose seems to be providing the opportunity for sound "effects,"[41] which served not in support of the narrative but as attractions themselves.

Manufacturers even developed a short-lived generic form based upon the illustrated song—the song film—which included, according to Altman, "His First Success" (Pathé, August 1907) about a cellist; "Dot Leedle German Band" (Kalem, September 1907); "Our Band Goes to the Competition" (Pathé, September 1907); "The Irresistible Piano" (Gaumont, November 1907); "The Merry Widow" (Kalem, January 1908); "The Mad Musician" (Selig, March 1908); "The Merry Widow Waltz Craze" (Edison, May 1908).[42] Edison released the picture songs "Love and War" and "The Astor Tramp" using several songs in sequence as well as stereopticon slides and reading matter. The Edison catalog of 1906 announced: "We have at last succeeded in perfectly synchronizing music and moving pictures. The following scenes are

very carefully chosen to fit the words and the songs, which have been especially composed for the pictures."[43] The film need not have been *designed* for
effects, however, since writers often praised drummers for the sophistication
of their "props" or "traps" and commended the tendency to look for every
opportunity to supply effects.[44] The emergence of sound practices that *asserted* their priority over the film, even if it had not been designed for sound
accompaniment, quickly became a central concern of manufacturers,
exhibitors, and critics.

Techniques of sound accompaniment were becoming so important to
exhibitors, and therefore to the emerging industry, that commentators
sought again and again to point to musical "effects" done properly, in order
to encourage "appropriate" practice. To be sure, in the 'teens and twenties
such effects were often ridiculed as excessive,[45] but for a time they represented one of the most sophisticated and lucrative approaches to exhibition.
More importantly, the use of effects and piano introduced a performance-
based aesthetic of sound to the average film and to the mainstream of cinematic practice. Like the barker, but in concert with the screen, the singer,
pianist, and drummer directly acknowledged their audiences, and even elicited their response through cleverness, ingenuity, and above all spot-on performances stressing synchronism. One measure of the sound practitioners'
success seemed to be exactness of his or her timing during performance.
These and other sound practices stressed performance to such an extent that
audience attention often seemed split between the world on the screen and
the performers in the theater. This dual focus fostered the tendency to evalu-
ate all sound, but particularly "synchronization," in terms of accurate performance.

SOUND AGAINST THE IMAGE

In addition to the stress sound effects accompaniment placed upon the performers, and therefore their direct and discursive relationship with the audience, it simultaneously and problematically altered the nature of the cinematic commodity. Although film producers had been generally successful
in consolidating their control over the product, sound accompaniment,
more and more a part of standard exhibitions after 1907, remained an arena
wherein individual exhibitors could alter the public's relation to the film.
Perhaps inadvertently, effects accompaniment introduced a punctuated
temporality of momentary audiovisual climaxes that could easily run

counter to the temporality of the more purely visual narrative. Moreover, such techniques could destroy the intended spatial hierarchies of the image. Several comic stories survive that describe overactive drummers and effects men who, in an attempt to "enhance" a picture, go to absurd lengths to punctuate every image with a sound. One writer describes lecturing with a film only to be interrupted by the repeated sound of a bird whistle. Perplexed as to its supposed source within the image, he asks the effects man and is pointed to "a diminutive canary in a tiny wooden cage on a top shelf at a far corner of the room," whereupon he "wallops" the whistler and proceeds.[46]

What makes this anecdote of more than passing interest is that it illustrates the conflict between two different conceptions of representation, and two modes of spectatorship. The drummer clearly searches the entirety of each image topographically in a quest for opportunities to display his craft, while the lecturer just as clearly adheres to what he feels is the self-evident hierarchy between narratively foregrounded elements and those left to the less important background. By adhering to a practice that was discursive and performative, the effects man *created* new hierarchies within the image, drawing spectator attention to incidental features because they could make noise. Under his gaze, the image ceased to signify in a predictable way, but became a pretext for virtuoso displays of sound. Consequently, the drummer could destroy the image's narrative legibility, or create discontinuities between shots where the producers clearly sought continuities.

The drummer's tendency to introduce potentially haphazard temporal and perceptual emphasis through sonic underlining presented producers with a dilemma: how do you control the type of sound performance in a theater in order to ensure the dominance of narrative over all other systems? The *Edison Kinetogram*, for one, began to publish musical suggestions for each film they advertised, as much to limit potentially transgressive forms of accompaniment as to suggest "appropriate" ones.[47] The drive toward standardization moved beyond music to encompass effects, too. A writer from the *New York Dramatic Mirror* suggests:

Incidental music is claiming the intelligent attention of some of the picture manufacturers, notably the Edison and the Vitagraph. Some time ago the Edison Company commenced printing programmes of instrumental music suitable for Edison releases, and recently, the Vitagraph Company announced that it would introduce properly arranged piano scores with each film of its manufacture. Now let some enterprising firm send along a prepared programme of sound effects to go with each subject, and another step forward will have been recorded.[48]

In general, the *Kinetogram*'s selections try to match narrative and musical mood, tempo, or theme. Falling within the category Bordwell calls "pleonastic" sound,[49] these selections were chosen in such a way that they were subservient to and supported the image-narrative. Each was chosen on the basis of its ability to mirror some important aspect of the story, rather than any other criterion, such as opportunities for virtuoso performative display or for audience sing-alongs.

Regardless of such attempts at standardization, sound still functioned as a mediating device between the impersonality of the mass-produced film and the particular screening situation. Although it is doubtful that many sound practitioners or musicians consistently achieved the kind of truly interactive give-and-take necessary for direct audience engagement, sound was nevertheless one aspect of the exhibition that, being discursive and performative, addressed audiences directly and offered the structural conditions for shaping audience response. As the institutional or industrial film tended to shift from an aggregation of "attractions" to a form stressing narrative integration, and while the represented space of the film became more and more distinct from the space of the audience, sound retained and continued to develop a mode of sonic "attractions" maintaining a direct relationship of performer/audience copresence.[50] Remaining one of the last arenas of exhibitor control, sound allowed various theaters not only to appeal to the religious, economic, and ethnic particularities of their audiences but offered the possibility of ridiculing the film, if it pleased the locals.

SOUND AND REALISM

Although performative sound implicitly threatened the dominance of the image, certain forms of sound practice could enhance both the exhibitor's and the manufacturer's prestige. Besides its highly touted benefits to attendance, sound apparently offered another type of symbolic capital that, itself, could be exploited to enhance revenues: realism.[51] Realism connected the realm of sound performance with the demands of narrative integration by producing the effect that sounds emanated from sources on the screen. Under the rubric of "realism," the drummer's sound effects were seen as adjuncts to the image and confirmed its own realism by creating the impression that sound emanated from it, just as in the world outside the theater. Such "realistic" effects participated in the more general historical trend toward greater integration of sound within the diegesis created by the image. Lecturers lectured less and imitated voices more, and drummers reined in their tendencies to show off in favor of supplying narratively plausible effects.

The situation of "realistic" sound accompaniment was more complex than I have suggested so far, however. Competing with the drummer was an array of "technologies"—Humanovo, Actologue, and Humanoscope, to name a few—that combined film projection with troupes of performers who provided voices and effects from behind the screen. True to the general trend, hiding the performers tried to downplay the discursive aspects of sound performance by subsuming them within the diegetic world, but contemporary descriptions nevertheless always comment on the preparation, execution, and diversity of performance.[52] The tension between realism and the flamboyant displays of skill it took to produce it profoundly shaped attitudes toward sound effects. Exhibitors such as Lyman Howe were cited for the care and practice with which they created their "illusion." Howe's career is exemplary of the importance of traveling companies to the history of sound in the cinema, as such groups were among the few who could repeat film material often enough to create a truly planned and practiced sound performance. A description of one of his shows is typical.

> The sound effects were not only in time and always appropriate, but their novelty and range is surprising. . . . There were such details in the pictures given sound as would never be thought of by any but a specialist; for instance, the silence of a Venetian Gondola ride is broken not only by the sound of the dipping oars, but even the rhythmic knocking between the oarlocks is heard. . . . [With a building's collapse] there was a single crashing, crackling sound equal in volume to the noise one would hear when watching the fire at such a distance as that at which the picture was taken.[53]

That the theater should be quiet enough to appreciate the dipping of oars is perhaps only as remarkable as the fact that such effects were attempted, and the reference to what might be called scale matching between image and sound is equally remarkable. As references to Howe—and especially his effects man, LeRoy Carleton—littering the trades confirm, it is consistently the quality of *performance* that was admired in his shows. Carleton, in particular, is praised as a consummate vocal performer, who "in the course of one

performance is called upon to make 115 changes of voice in his vocal mimic-
ry of both humans and animals."[54] One reviewer argued that

> the 'noise' portion of the show—the use of stage effects to make the pic-
> tures more like real—is the best that has ever been used in Pittsburgh.
> Conversations of the subjects of the pictures, expressing every emotion as
> depicted on the faces of the pictures: the whirr of machinery, rumble of
> railroad trains, swish of water in marine scenes, and various other things
> that help the onlooker to imagine that he is witnessing the real thing
> instead of a counterfeit presentment.[55]

Such "realism" crucially depends upon precise performance, and is therefore
always capable of usurping the film. Altman suggests, however, that using the
voice in this way decisively tipped the balance away from the discursive tradition of the lecture and made speech "another sound effect, a method of further anchoring the film experience in reality" by solidifying the effect of simulated presence.[56]

SOUND IN THE SERVICE OF THE IMAGE

Even while drawing attention to the performer's skill, sound effects beyond
the realistic use of the voice could still enhance the representation of narrative rather than deflect it. Indeed, sound accompaniment could become
essential to the narrative film. One writer commenting in early 1908 gives an
example of a film whose meaning was dramatically clarified by the intelligent use of effects. He enters a nickelodeon to view a film in which a couple,
while crossing a field, suddenly pause and, inexplicably, bow their heads
before proceeding. Seeing the film again in another theater, he notes that the
previously puzzling pause is accompanied by the sound of a church bell,
which clarifies the meaning of the gesture immeasurably.[57] In this case,
sound *produces* the film's legibility as a unified narrative, motivating what
would otherwise be a perplexing discontinuity. This technique served the
silent film long and well, and a similar effect is used early in Vitaphone's *A
Woman of Affairs* (1928), where several "silent" characters react to the sound
of a horn offscreen. Had the film been projected without sound (as was likely in some unconverted theaters), the motivation of the characters' response
would have remained slightly obscure, but with a synchronized effect it is
perfectly understood.

This sort of clarification became a more and more pressing issue between 1907 and 1910, as plots became longer and more complex and demanded an explicit narration. Some writers accurately pointed out that, for example, a version of *Macbeth* (1908) reduced to a single reel, or the various "Films d'Art" lately arriving in the United States, simply *required* the presence of a lecturer, lest they remain hopelessly opaque.[58] Earlier condensations like Edwin S. Porter's *Uncle Tom's Cabin* (1903) relied to a much greater extent on the spectator's presumed familiarity with the story and consequent ability to fill in the gaps or discontinuities. Typical of the industry-wide trend toward self-contained modes of narration, however, these later films did not assume the same audience foreknowledge, and elliptical narratives became a problem rather than a strategy.

While some writers suggested that lectures simply helped the "less intelligent" understand the increasingly complex stories,[59] others like Van C. Lee championed the film lecturer, not as an attraction himself but as a necessary adjunct to the image, arguing that lectures can help everyone understand multishot films better. Like other writers promoting the "intelligent" use of sound effects, he cites Lyman Howe's considerable success as an example of the lecture's benefits.[60] One writer even argued that motion picture producers were so concerned with the problem of intelligibility that they meant exhibitors to read the film descriptions they included with the rented film aloud during screenings. This writer also noted that only a tiny minority of exhibitors ever did so.[61]

In contrast to those who praised the promiscuous use of sound effects, other writers began to insist upon limiting them to "appropriate" effects. Even the drummer/piano pair praised for their tendency to "take every opportunity for effects," and to "match all movements," indicate that they, nevertheless, always play "with" the picture.[62] Other writers concur, arguing that, above all, musicians should avoid "interpreting" the picture wrongly. Thus, some argued even the ubiquitous coconut shell hoofbeats precisely synchronized to the image need to change their tone when the horse moves from pavement to dirt, lest the clip-clopping on a marble slab become comic.[63] Noted commentator W. Stephen Bush sums up the general tenor of the discussion, praising the voice that "runs with" the picture, explaining the figures and plot while bringing out "by sound and language the beauties that appear but darkly or not at all until the ear helps the eye."[64]

To a remarkable extent, Bush's comments illustrate the growing tendency of one tradition of sound to absorb and integrate another that, otherwise, appeared fundamentally opposed to it. The "performative" tradition, allied to a more topographic and less narrative approach to the image, tended to treat the image as a pretext for the gratuitous production of sound. Deriving its impetus from vaudeville, it stressed comic accentuation and an intermittent, punctuated temporality, resulting in an antinarrative and antipsychological form of humorous attention-grabbing. The later, and ever more dominant, tradition stressed a rigid hierarchy in providing sound effects, separating the image into zones of importance and of unimportance. A gun wielded by the villain was an appropriate cue for sound effects, but a tiny background canary was not. Above all, synchronized sounds had to clarify or underscore the story by matching mood, tempo, or character psychology, and simultaneously create and enforce the hierarchies of the image. Like so much else discussed here, sound effects hierarchies were ultimately determined by the necessities of the increasingly complex, psychologically motivated, multishot narrative film, which had emerged as the commodity upon which the emerging industry was erected.

As these trends indicate, one of the principal techniques that classical filmic narration uses to create continuities out of the heterogeneity of formal elements is the motivation of discontinuities, such as produced by the material interruptions of the multishot image track. The proper "motivation" of these evident discontinuities results in a unity at another level, that of the narrative itself.[65] A clearly hierarchical visual space where narratively essential elements (say, human characters) are foregrounded in visual space against a generic background allows for an easier link to succeeding shots, since the spectators are already "primed" to attend to those characters. Images that lack such a hierarchy (no distinction between "important" and "unimportant" elements, say) require that each shot in a series be scanned in its entirety, topographically, and therefore with less emphasis on continuity *across* edits.[66]

Therefore, any form of sound accompaniment that threatened to disrupt the priorities of image space by redirecting attention to narratively inessential elements similarly threatened the norms of the hierarchical and linearized images. Increasingly, all forms of sound work were required to become, loosely speaking, pleonastic, underscoring and highlighting an image hierarchy that was, in Hollywood, always mandated by the generative narration. While it remained possible for a sound practitioner to display skill in synchronizing effects, the range of permissible sources for sound was ordered by the demands of the functions preordained for them by the narrative. The villain's gun could always make a noise, because it was always important, but individual guns in a battle used as a backdrop only merited a generic matching. However, the rolling surf in the background of a scene could get a prominent sound accompaniment provided it did not interfere with the legibility of the foreground narrative, or if the surf itself, for narrative reasons, was being "foregrounded." As I will argue below, much the same series of accommodations were required of musical accompaniment. Summarizing what would become the accepted wisdom of the post-1910 era, S. L. "Roxy" Rothapfel counsels that "pianists must remember that they are merely one of the cogs in the wheel that makes the picture theatre go round, and as a rule, the people pay to *see the picture, not to hear the pianist*, so therefore, play softly when occasion demands, and always remember *the picture comes first*."[67]

"SYNCHRONIZING" THE PICTURE

The modes of audience address characteristic of film music were occasionally at odds with the typical address of the image track. The richest, longest-lasting, and most familiar tradition of synchronization (as it was called in the silent era)[68] involved "fitting" music to the picture. Influential how-to books by Robert Beynon, Edith Lang and George West, and Erno Rapée discuss the various methods of "synchronizing" music to a film.[69] As implied by the *Edison Kinetogram*'s earlier musical suggestions, these writers indicate that the so-called compiled score, fashioned from existing selections chosen to match the scenes in the film, dominated musical accompaniment. The two other major approaches, the original score and the leitmotiv principle, while practiced by some, were comparatively rare. The reasons for this are several. Among other things, compilations relied on well-known but not copyrighted material, thus requiring less new learning and skill to play. Unless written in advance by a composer, the improvisational use of leitmotivs in a manner that did not simply repeat the same themes ad nauseam was beyond the means of almost all musicians.[70] As the true leitmotiv system could not, frankly, be improvised in any case, compiled scores could still associate character or situation with a particular melody. Remember, too, that these were often unschooled and vaudeville-trained musicians skilled in accompanying hand springs with cymbal crashes and punch lines with rim-shots. Although it is certain that something *like* the leitmotiv system was often used to underscore narrative relationships, this tendency did not by any means dominate, nor did its compatability with the character-centered narratives of the emerging classical cinema somehow define, other musical practices as beyond the classical paradigm.

Given the number of factors militating against the widespread adoption of the original score and/or leitmotiv system, it is surprising that other methods of "synchronizing" sound to a film have not been given more attention in the nonmusical histories of the era, especially since they help to explain some important aspects of film form.[71] As the Edison films "Love and War" and "The Astor Tramp," which chose images "to fit the words and the songs,"[72] attest, it became common practice to match film scenes with musical selections whose title, style, or formal qualities were somehow related. Sometimes they matched in tempo, sometimes in mood, sometimes rhythm, sometimes in what they seemed to depict, but often they did so without regard for the connection *between* scenes.[73] It has even been suggested that cases such as these had a precedent in the *opera pasticcio*, whose libretto provided the context and the "excuse" for connecting an assortment of otherwise unrelated "greatest hits" from existing operas.[74] As in a musical revue, the narrative exists for the purpose of motivating a series of show-stopping numbers exhibiting something like what Noël Burch calls "primitive autarky."[75] Again, like some forms of sound effects accompaniment and like the earlier song-film, the image-based narrative is subservient to the direct-address of the performance, eliciting praise for the musicians rather than the film.

Musical accompaniment that emphasized discursive relations over submission to the film included the practice of "funning" a picture. "Funning" satirized scenes or entire films through musical puns, commenting on the picture through the title, lyrics, or melody of the accompanying music. A *Moving Picture World* columnist pointed, for example, to the use of the song "Oh You Kid" in a scene where the pharaoh's daughter discovers the baby Moses.[76] Another writer argued, "Bad judgement in the selection of music may ruin an exhibition as much as a good programme may help it. Imagine a pathetic scene showing a husband mourning his dead wife accompanied by the strains of 'No Wedding Bells for Me!' And yet this exact circumstance was noted by the writer recently."[77] Such techniques catered to particular audiences and their prejudices and/or drew attention to the musician's cleverness or stupidity, but did so to the detriment of the film, whose uniformly coherent address was hopelessly fractured.

In conflict with the trend toward regularizing musical accompaniment and robbing it of any nonsanctioned voice, many authorities, including Beynon, still considered a necessary component of good sound work to cater to the ethnic, class, and religious particularities of the audience.[78] Funning fell as quickly out of favor with music columnists and large theater orchestras in the 'teens and twenties as did the unreflective use of musical clichés like

"Home on the Range" and "Rock-a-Bye Baby" for the all-too-obvious scenes. But in small houses in ethnic enclaves or theaters with a single pianist or organist, these techniques continued to provide a nonnarrative or meta-narrative pleasure while implicitly cementing a bond between performer and audience. To the consternation of manufacturers, this community often came together at the expense of the film.

Theater owners learned quickly how to improve receipts by catering to local audiences through attention to particular ethnic or cultural tastes and traditions. Sometimes this attention amounted to little more than not programming a boxing match along with a Passion play, in deference to the more devout audience one could expect for the latter. In other cases it involved hiring conservatory-trained pianists to cater to Russian, Romanian, and Hungarian Jews who were thought to be "hard to please, musically."[79] Other writers, however, criticized sound practitioners who spoke in slang, with accents, or ungrammatically because these performers supposedly showed less respect for the audience. Patrons who were supposedly "deceived" into believing that they had attended a genuine (phonographic) talking picture had their illusions shattered by a writer for the *Moving Picture World* who heard nothing but "Dis, Dat, and Dem," coming from behind the screen.[80] Of course, this argument was a double-edged sword since one imagines that these mispronunciations unmasked the ruse because they were too much like those of the audience, and not the expected anonymous or "correct" type. Such grammar and pronunciation, and ethnically specific address in particular, were discouraged in the name of attracting a "better" crowd, yet what may have seemed "imperfections" in grammar from one point of view may, from another, have kept local audiences coming back.

Ultimately, the mass-produced and mass-consumed nature of the movies obviated such specific forms of address as it simultaneously sought to eradicate those forms of spectatorship that emphasized communal, collective, and local responses. Instead, they sought forms that encouraged an isolated, individual, and silent audience—that is, the norms of middle-class theatrical spectatorship. Despite these powerful trends, alternative modes of presentation and spectatorship persisted well into the late silent era, as scholars such as Mary Carbine have shown.[81] Theaters in Chicago's "Black Metropolis" employed jazz bands who, apparently, improvised along with Hollywood films, often adding "hot" numbers where Hollywood would have desired pseudoromantic mood music. One can easily imagine an orchestra with players like Louis Armstrong taking great delight in elaborate musical puns played at the film's expense, and given the enormous improvising talents these bands must have included, all sorts of "funning" from sound effects jokes to race-specific "commentary" must have been common.[82]

In contrast to such techniques, books like Rapée's *Motion Picture Moods* (1924) and the various collections published by G. Schirmer and Sons encouraged pleonastic and supportive "isomorphisms" between narrative and music. Melody, with its necessary beginning, middle, and end, could mimic the structure and unity of the plot. The climax of the music, they advised, should match the climax of the narrative; fast-paced scenes should be paired with fast-paced music, battle scenes with battle-like program music, and love scenes with romantic melodies, thereby formalizing the growing trend toward redundancy of sensory address and information. Although the appropriate selections for each scene were still discrete and often tonally incompatible, these "how-tos" offered advice and examples illustrating the importance of modulation techniques that could provide continuous musical bridges between themes in different keys. These bridges functioned in a manner that imitated the image's emphasis on creating a continuous whole out of discrete fragments. These essentially improvisational techniques allowed musicians to shorten or lengthen selections without obvious interruption, and to create a form of narrative integration between separate elements. In fact, some writers came to favor the accompaniment of the piano or organ over the orchestra precisely because it was easier for a single well-trained and flexible player to improvise connections between a piece in F and another in C minor.[83]

In addition to mimicking the pluripunctual unities of the image—spatiotemporal wholes constructed across a series of discrete shots—Beynon, and Lang and West, counseled would-be pianists and organists to make themselves and their music subservient to the needs of the plot as well. While Beynon, for example, suggests that it is still appropriate to cater to the "religious, social or political leanings of a neighborhood," or to make a pun,[84] these should be strictly limited since, in excess, they may distract from the picture. Beynon is adamant when he warns, *"Never portray musically an emotion contrary to that depicted by the screen action."*[85] Lang and West provide a similar warning when they assert that "the player must be careful not to 'italicize' the situation so that it becomes distorted or burlesqued. Therefore he should avoid all excesses."[86] Excess is defined in strictly narrative terms as what exceeds the demands of plot, emotional consistency, or tonal coherence. Both books argue that the ultimate goal is not to attract attention to the music, since this would "detract from the smoothness necessary in good pic-

ture accompaniment, and the leader should regard smoothness as of paramount importance."[87] In addition, they argue against the frequent use of
popular songs on the grounds that they may be overfamiliar or encourage
audience members to whistle in the theaters. They likewise argue against folk
songs, as they are often associated by the audience member with "some crisis in their careers," and against all national anthems except as a form of narrative redundancy.[88]

The general principle unifying these various suggestions encouraged
musicians to minimize all those aspects of the presentation that could pull
spectators out of their diegetic absorption, drawing attention toward the performers or other audience members and away from the image narrative.
Likewise, the new attitudes implicitly argue against too-specific understandings of the audience and too-local means of addressing them and their concerns. All the advice offered by both books assumes the norms of middle-
class spectatorship and specifically warns against encouraging any "deviant"
spectatorship, whether through sing-alongs, burlesque humor, or parodies of
the picture based on local attitudes toward the events or people depicted in
the film. Ideally, the accompanist should be able to walk into a theater any-
where in the United States, play along with the picture, and please his (nonspecific) audience.

In what, then, did the vaunted "smoothness" consist? In essence, it seemed
to require a kind of musical "narrative integration." Like the images it
accompanied, silent film music moved from a mode of intermittent "attractions," or a mode stressing the autonomy of individual elements, to a form
that stressed the absolute continuity of the music from one end of the picture
to another.[89] Given the standard techniques and predictably variable quality
of musical accompaniment, this was not always such an easy task. As the
'teens progressed, writers like Clarence Sinn and Clyde Martin argued for the
new norm of uninterrupted, continuous musical accompaniment, from the
beginning of the picture to the end.[90] While Lang and West argue for an
approach that starts from a memorable theme or two and builds the entire
musical structure from improvisations around them, Beynon, writing only a
year later, believes this technique encourages too much "key twiddling." He
argues instead for a more carefully prepared approach, centered not on
improvisation but on the construction of a written score.[91] Beynon summa-
rizes the practices of music "synchronization" as follows:

> The secret of synchrony lies not so much in careful timing of the selections
> as in the accurate judgment of the musical director. Music need not be cut

to fit the situation; but, if care be taken in the finishing of phrases, the musical setting becomes cohesive—one complete whole that conveys to the audience that sense of unity so essential to plot portrayal.[92]

In other words, musical accompaniment should produce precisely the sense of integrated and satisfying closure as the plot. Constructing such a musical plot, however, was a fairly complicated process.

Beynon splits scores into three groups—the original, the compiled, and the semioriginal. Since original scores, while now the norm, were comparatively rare in the silent era (being confined, for the most part, to prestige features in large urban houses), Beynon (and to a certain extent, Lang and West) bases his general musical approach on the "semi-original" score, developed to rectify "the faults of the compiled score."[93] The table of contents of one of Erno Rapée's books (*Motion Picture Moods*), which collect bits of existing music useful to a motion picture organist, can give a sense of the compiled score. Rapée divides the material into categories like Aeroplanes, Battles, Birds, Dances, Fire Fighting, Gruesome, Lullaby, Monotony, National, Neutral, Oriental, Railroad, Sea and Storm, and Wedding, using pace, mood, emotion, nation, events, sources, and other categories as equivalently descriptive.[94] A score cobbled together out of a list such as this could easily collapse into dozens of disconnected fragments, as the pianist shifted from "Lullaby" when a baby appeared on the screen, to "Gruesome" as the villain approached, to "Race" as the father raced home to the rescue. The potential for discontinuity was, without a doubt, the peril that musicians were taught to avoid most. Like the primitive "mode of attractions," or a "mode of discontinuity" that left individual textual units their autonomy, the compiled score, even when it matched the individual shot brilliantly, could work against the hard-won narrative continuity of the edited scene. Despite the obvious necessity of the process, Rapée makes no mention of how these selections were to be combined.

In general, these authors promote a specifically psychological and narrational musical aesthetic that requires the accompanist to "fit the predominant emotion," discover the "significant nucleus around which the silent drama is built," or locate "the dominant."[95] He must be "emotional," needs "psychological insight," should recognize emotions in the characters, and "above all, learn to read facial expressions."[96] As if teaching their readers to write scenarios, they clearly describe the Hollywood film as a character-based, psychologically motivated, linear narrative form. Given that linearity, smoothness, and coherence are the musical goals, as they are for the picture, we might expect them to encounter some of the problems characteristic of the multiunit image-narrative.

In contrast to Rapée, Beynon as well as Lang and West spend much of their books teaching techniques of modulation, or improvised key-changing, in order to connect the various selections into a more or less continuous whole. Specifying their understanding of film continuity, Beynon stresses upper-voice melody, and Lang and West a "solo-instrument-with-accompaniment" approach, in order to achieve the musical equivalent of the foreground/background structure so essential to the narrative film image.[97] Keeping the dominant melody audible by placing it in an upper register or by giving it a different "voice" on the organ is not so different from staging the action so that the characters are centered and in focus, in order that we may follow them from shot to shot. Continuous melody, therefore, assumes a role analogous to the image foreground and is susceptible to its articulations. By this logic, the melody would change when scenes changed or when locations shifted decisively.

The manner in which the authors handle cross-cutting and multishot scenes involving inserts or, as they were known at the time, "flashbacks" is very instructive for understanding how musical and filmic continuity were defined at the time. Given the implied link between melody and neoclassical unities of time, place, and action, the flashback posed a particularly tricky puzzle. As a result, each writer warns against mechanically mimicking the material discontinuity of the image (the succession of shots) by changing the accompanying music with it. Beynon argues that the musician should fit the predominant emotion of a scene and advises against changing the theme with each flashback or insert shot (allowing that the tempo might be changed, if not the melody). This strategy attempts to avoid the effect of choppiness in what, after all, are not separate scenes but only separate shots. Lang and West similarly argue that during a "flash" (i.e., an insert, mental image, another location, etc.) the music should not change its motive. Each recognizes that the material discontinuity of the image is merely a change in narrational level, "therefore the music should not change its *character* during a flash-back, but it should be very much *subdued*. . . . The principles are: in most cases *not* to *disrupt* the continuity of the music while the flash-back lasts, but to *change* the *intensity* by playing the music, characteristic of the main action, in a dynamic degree of loudness or softness which befits the secondary action."[98] The authors add that in a cross-cut scene, however, a "competition" between themes might very well be appropriate.[99] Rapée makes identical recommendations regarding volume and melodic "competition" in his later *Encyclopedia of Music for Pictures* (1925).[100]

These musical recommendations indicate a great deal of sensitivity to the intricacies of Hollywood narration, and are so flexible that they can be applied to a much later film like *Old San Francisco* (1927), where scenes between two musically delineated characters are accompanied by the theme of the one who dominates the scene. As in related suggestions requiring musicians to vary tempi according to the plot, or volume according to the proximity of the narrative climax, here the authors recapitulate many of the principles by which the Hollywood narrative is constructed, further submerging music within the demands of classical narration. Despite all the possibilities for audience address available to musicians, the address of the fictional image-narrative is absolutely sovereign for these writers, and would become almost inevitably so for all musicians in the sound era.

THE PERSISTENCE OF MARGINAL FORMS

Pre-talkie sound was, it is clear, a multiple and heterogeneous set of phenomena comprising a wide variety of functions. From "barking" to hailing, within and outside the text, sound emphasized its presence and its necessity by organizing and reorganizing the cultural practice of cinema. Sound often placed an enormous emphasis on the status of cinema as a kind of performance, whether through the elaborate practices of synchronous sound accompaniment, lecturing, or musical presentations of various kinds. Before 1910, sound very often dominated or threatened to dominate the image, in terms of conception, production, presentation, and even reception. Sound-based genres like the travel lecture and illustrated song, the visualized record, or the film that functioned as a pretext for musical or effects performance occupied a prominent place in the world of early cinema. As the image-based narrative began to assert its dominance, however, sound's functions began to change. Sound could serve as a clarifying addition, as in the lecture, or it could provide important narrative motivation for on-screen events by supplying a missing sound cue. Or, musically, sound could aid in characterization; it could underline emotions, set moods, or emphasize tempi. On the other hand, music and commentary could work against narrative unity—by punning, by meta-commentary on the situations or characters, by irony, all of which might serve as a way to mediate between the film's anonymous, mass-cultural address and its ethnic, economic, and social particularities.

While, in general, sound (especially music) did tend to become more and more subservient to the narrative, none of these traditions of sound disappeared entirely; they persisted on the margins of the classical style or in alternative modes such as the avant-garde. The role or roles that recorded sound was to fulfill were therefore anything but self-evident and would certainly be (as I hope I have suggested) the object of struggles for years to come. Simply because the transition to sound was relatively smooth does not mean that we should ignore the processes and principles of exclusion by which the classical style operates and defines itself and its boundaries.

Despite the growing symmetry between the form and address of the image-narrative and that of sound, even in studio-composed original scores, important elements of another less fully narrative mode persisted. As the *American Cinematographer* noted in 1928:

> The screen drama of the immediate future will involve, not conflicting character, but competitive noise. The exposition of the rural drama will present the problem: "Who will emerge victor in a hog-calling contest and win the girl." The development will proceed from one minor crisis and crash to another with steadily increasing volume until the heavy, with a portable boiler shop, threatens to drown out the hero's calliope. The denouement will come when the heroine blows up the field of oil tanks with her cigarette and scatters fragments of the heavy all over the pleasant meadow.[101]

As sarcastic as this seems, the writer *was* fairly accurate. Vitaphone's 1927 *Old San Francisco*, otherwise a standard "silent" melodrama with silent film sound conventions based on character and ethnic difference, ends with a spectacular earthquake lasting several (ca. eight) minutes chock-full of synchronized sound effects. Just as the character O'Shaughnessy's Irishness is motivated less by the inherent demands of the narrative than by the opportunity it offers to differentiate him by musical theme and written titles, so the earthquake's spectacle is unwarranted by the film's narrative economy. Earlier parts of the film are sporadically accompanied by diegetic sound effects often tied to dramatic narrative points. There is even an audio "point-of-view" shot showing, first, someone cocking an ear, then a close-up of horses' hooves. The earthquake sequence, however, contains the film's overwhelming majority of sound effects, and does so in an excessive manner.

The nonstop symphony of crumbling buildings, running victims, roaring fires, and Anna May Wong's climactic scream offer a density of effects unheard in silent film, yet not entirely without precedent. Like some predecessors, this spectacularization of sound was not primarily narrative in motivation, nor essentially "realist." Indeed, it stalls the plot, and its length is narratively unjustified. It does, however, foreground the *technology* of sound recording and sound reproduction, offering its perfect performance for appreciation and admiration. The sound does not support the narrative here, but narrative supports the gratuitous display of sound. Likewise, in *The First Auto* (1927), synchronized effects are not primarily narrative. From the opening drumbeat hooves and the shout of "C'mon Hank" (over an image of a crowd), *which is followed by an intertitle of the same phrase*, through the sudden eruptions of a synchronized "Bob!" which is followed by conversation on intertitles, to the gavel that starts the auction, sound is foregrounded not in order to present narratively important information (else it would not repeat "C'mon Hank") but to address the audience directly—to hail them—to say, "Hey! Look and listen! This is important." These visible sync-events serve to foreground the spectacle of technology while disregarding narrative.

Interestingly, the idea of sound, and especially synchronization, as *performance* carries over into the sound era, and into the realm of technology, as the ability of the device to ensure accurate synchronization seems foregrounded, itself, as a kind of performance.[102] The earliest patrons of the Vitaphone certainly must have appreciated the new technology along these lines. *Don Juan* (1926), *The First Auto*, and *Old San Francisco* offered numerous opportunities for a spectacularization of the device's performance through foregrounded sync-events. At points, the stories seem to exist purely to provide an excuse for a display of effects, in a manner that recalls the excuses for dramatic lighting effects in many DeMille productions.[103] However important narrative may be in these films, at times it clearly takes a backseat to the display of pure technological marvel, and cinematic pleasure momentarily divorces itself from plot.

Nor did the "sound over image" tendency die with the silent cinema. The illustrated song, in particular, persists today in the form of music videos, for example, as it did in the thirties. A cartoon like Betty Boop's *Snow White* (1933), featuring Cab Calloway, included songs owned by Paramount, its distribution company. Given the structural primacy of the sound track in these song-films,[104] it is easy to see how they descend from the illustrated song. In a similar fashion, *The Big Broadcast* (1932) (also by Paramount, and featuring Cab Calloway) exploits several of the same songs as the Boop film, and even has an ingenious opening segment which mimics, with "real-world" images, the isomorphic techniques of the cartoon's "mickey-mousing."[105] Again, although this unusual sound/image strategy gets a certain amount of narrative justification (it takes place in the clock-obsessed world of a radio studio), its chief motivation derives from the economic and structural importance of the sound track.

Although the "isomorphisms" of musical accompaniment and of the cartoon seem fully a part of the Hollywood narrational system, isomorphic techniques also supported one of the strongest nonclassical traditions of film sound. The basic principles of synchronization, while certainly amenable to the classical film and capable of clarifying its images, were by no means confined to it. From Dziga Vertov's *Enthusiasm* (1931) and Eisenstein's *Alexander Nevsky* (1938), through mickey-moused cartoons, to Peter Kubelka's *Unsere Afrikareise* (1961–1966), and Larry Gottheim's *Mouches Volantes* (1976) and *Four Shadows* (1978), a marginal but powerful tradition of synchronization assumes the foreground. The sound and image relationships in these films are among the most carefully "synchronized" in the history of film, but they work in a manner that, while partially applicable to the Hollywood narrative, is generally quite opposed to it. In contrast to the more familiar sense of the word *synchronization*, in these films both sound and image "articulate" one another, ignoring the conventional divisions and hierarchies appropriate to narrative. As we have seen, however, traditonal, narrative synchronization may be a "special case" of this more general principle.

A PREORDAINED ROLE FOR SOUND?

As I've suggested, not only did Hollywood waver in its allegiance to representational modes during the transition, moving between discursive and diegetic forms, but the mode of representation understood to be characteristic of the very technology of sound was likewise up for grabs. Was sound an effect? Was it narration? Was it clarifying commentary? Should it function as an added form of omniscient or restricted narration? Was it realist or spectacular in nature? Even its technical nature was in dispute. Was it closer in form and purpose to the phonograph, the telephone, or the radio? Each device, while useful for grasping some aspects of the sound film phenomenon, validated different techniques and implied different representational norms. So, before "sound as sound" could be "inserted into . . . the Hollywood style," it had to be determined just what sound "was" and what its appropriate functions were.

Recent histories have emphasized the ease with which sound was adopted into the classical paradigm, deafening and blinding us to the very real discontinuities of the transition and closing off the possibility of a history based on the tension between competing norms.[106] Ideas of technological and representational self-evidence and self-identity and notions of functional equivalence derived from the end point of a historical transformation encourage us to hear history as a monologue. Concepts like jurisdictional struggle, technological redefinition, and functional near-equivalence[107] provide the contradiction and conflict that can motivate historical change, encouraging us to listen to quieter voices from the past.

CHAPTER 4

SOUND THEORY

While the basic issues I have identified as arising from the emergence of the technical media are in some ways of quite general import, the manner in which any particular representational technology is normalized, regularized, and institutionalized is always quite specific. Just as the nature of Marey's and Helmholtz's experiments placed specific representational demands on contemporary sensory devices in order to render them adequate to the rigors of science, the emerging phonography and film industries asserted their own requirements for the proper implementation of both camera and sound recorder, shaping their characteristic uses in thoroughgoing ways. Contrary to what we might expect, explicitly theoretical debates about the general nature of technological representation were central to the development of Hollywood representational norms, and helped determine how sound technologies were understood and deployed. Particularly at moments of institutional crisis, reorganization, and transformation, these debates assumed a discursive importance that often shaped technological research and aesthetic experimentation, serving both as practical standards and regulative ideals.

In this chapter I will explore several influential manifestations of sound theory, examining them not only for their internal coherence and insight but also for the role they play in determining different forms of representational practice. As I have suggested in previous chapters, I am not only interested in how explicitly theoretical arguments illuminate our understanding of aes-

thetic form and pactice, but perhaps even more in how representational practice can often imply, embody, or even produce a sophisticated theoretical
knowledge of its own. Marey's systematic experimental interventions in the
realms of performance, inscription, and reproduction, like Edison's similar
discoveries with the phonograph, can be mined productively for their practical (and unstated) theorization of the processes of technological representation. In a parallel fashion, this chapter examines the practices and theories of
sound technicians during Hollywood's transition to sound film production. I
also examine academic theories of sound recording, which address in self-
conscious and logical fashion the issues that technicians raise in the realm of
practice. In effect, I look at two kinds of theory, each of which finds it roots
in the nineteenth-century encounters with technology, and each of which
confronts the inscription/simulation dichotomy in a characteristically different way.

ORIGINAL/COPY: SOUND THEORY?

As we have seen, from the early days of this century those engaged in sound
recording as a practice—as a business, if you will—recognized its many representational dimensions and complexities. By contrast, the theoretical discussion of sound within film studies has returned again and again, almost
obsessively, to a single central problem. What is the relationship between a
sound recording and the sound it purports to depict? Béla Balázs, for instance,
tells us that "there is no difference in dimension and reality between the original sound and the recorded and reproduced sound." Stanley Cavell tells us
first that "in a photograph, the original is as present as it ever was," and then
derives the claim that "sound can be perfectly copied . . . the record reproduces the sound." Jean-Louis Baudry tells us that "one does not hear an image
of the sounds, but the sounds themselves . . . they are reproduced and not
copied," and Christian Metz that "auditory aspects, provided that the recording is well done, undergo no appreciable loss in relation to the corresponding
sound in the real world: in principle, nothing distinguishes a gunshot heard in
a film from a gunshot heard in the street." Finally, Gerald Mast asserts that
"there is no ontological difference between hearing a violin in a concert hall
and hearing it on a sound track in a movie theater."[1]
    More recently there has emerged an opposed group, all convinced, like
Alan Williams, that "it is never the literal, original 'sound' that is reproduced
in the recording, but one perspective on it, a sample, a reading of it." Rick

Altman tells us that recordings have "only partial correspondence to the original event," and that "recorded sound creates an illusion of presence while
constituting a new version of the sound that actually transpired." Most forcefully, Thomas Levin asserts that "familiarity [with recorded sound] has dulled
the capacity to recognize the violence done to sound by recording."[2] Despite
minor divergences, all three insist on the fundamental importance of presuming nonidentity between original and copy.

Williams, Altman, and Levin build their positions on the careful and very
specific definition of basic categories, the most important being sound itself.
Williams provides the thesis from which all three work: " 'sound' . . . will be
taken to mean 'audible disturbances of air in the form of wave motion in a particular configuration of space.' "[3] Thus defined, sound is a "three-dimensional,
material event," whose inseparability from the environment of its production
necessarily entails the consequence that "a sound recording can not by definition reproduce the 'sounds themselves'—since it is obligated by its nature to
render a sound (as vibrating volume) as recorded *from one point* of the space in
which and through which the sound exists."[4] Inseparable from the time and
space of its production, each sound becomes an essentially unrepeatable *event*
distinguishable from all others.[5] As Levin puts it: "If a sound is understood as a
volume of vibrating air, then assuming for a moment that sound reproduction
were absolutely flawless . . . such a 'perfect' reproduction of sound waves in a
different volume would effectively constitute a *different* sound."[6]

These "nonidentity" theorists agree that original sounds are unique and
inherently unrepeatable, and in an important corollary Williams notes that
even the original itself is intrinsically multiple and internally differentiated—
a fact we recognize every time we choose between "good" and "bad" seats in
an auditorium.[7] Every recording must, perforce, select a point from which to
"hear" an event, thus,

> in sound recording, as in image recording, the apparatus performs a signif-
> icant perceptual work *for us*—isolating, intensifying, analyzing sonic and
> visual material. It gives an implied physical perspective on image or sound
> source, though not the full, material context of everyday vision or hearing,
> but *the signs of* such a physical situation. We do not hear, we are heard.
> More than that: we accept the machine as organism, and its "attitudes" as
> our own.[8]

By modeling his own critique on Jean-Louis Baudry's essay, "Ideological
Effects of the Basic Cinematic Apparatus," Williams limns a sonic version of

apparatus theory and provides a foundation for an analysis and critique of the ideological work of sound recording—work that Balázs, Baudry, Mast, and Metz seem unable to identify, let alone critique. Levin, whose argument with Metz is the most sharply focused, describes the stakes of their disagreement in the following terms:

> That a gunshot *seems* to sound the same in the different acoustic spaces of the street and the inside of the cinema *is a deception.* If differences remain unnoticed, this is a function of a socially constructed auditory practice which emphasizes the similarity of such sounds in order that they be understood (i.e., linked to a common source) by the hearer.[9]

While in light of these criticisms, any claim of identity between a sound and its representation must seem profoundly naive, Metz's claim that "nothing distinguishes a gunshot heard in a film from a gunshot heard in the street" is anything but. Instead, it evinces rather different understandings of film sound, of audition, and of sound in general.[10]

Metz is surely aware that the experience of hearing a gunshot in the street hardly resembles hearing one at the movies. In fact, as Altman points out, they need resemble each other in one respect only—in name.[11] One cannot even say that the *source* of the sounds is the same, since there are dozens of acceptable substitutes for a movie gunshot—many of them more effective than the real thing. As Metz implies, in a *functional* or *narrative* sense what matters is our ability to identify a sound's source, and thereby its meaning. In other words, Metz understands sound events to be inherently *legible*, or recognizable, across a series of different contexts as relatively stable signifying units. Against the definition of sound as an essentially unrepeatable *event*, Metz describes sound as an eminently repeatable and intelligible *structure*. Metz, in effect, rejects the current version of representation-as-sensory-simulation in favor of a new version of representation-as-(legible)-inscription. Metz thereby emphasizes less its perceptual uniqueness than its capacity to generate meaning in a particular context, defined in this case by economic, institutional, and formal parameters.[12] For Metz, it is not a question of there being no *literal* difference between street and theater, but rather there being no difference in *meaning*.

In order to address problems in theories that stress the identity of original and copy, Levin argues for a "critical analysis" of the sound apparatus "to understand how sound [like the image] . . . is transformed in the process of its reproduction."[13] He maintains that it is the critic's task to reawaken the "capacity to recognize the violence done to sound by recording." However important, it seems ill-advised to ground that analysis on the assumption that there exists an "original" sound that is pure and present, that remains inviolate or wholly prior to the representational act's "transformations," against which to measure all sound recording. While Williams and Altman warn us against approaches that reduce sound *representation* exclusively to sound *reproduction*,[14] they, too, nevertheless assume a standard, and essentially *prior* sound, which is transformed or violated by recording—a sound that is "partial," "pre-digested," "blunted," "sampled," "neutralized," and "homogenized" by the technology "itself," and necessarily by *any* application of the technology.[15] Consequently, each also locates whatever ideological significance there might be in the recording process in the difference between original and copy, and thereby defines that relationship as singularly determinant. Ironically, Edison, who seems never to have had a single theoretical thought on this topic, already knew better, and ignored the original/copy distinction except when it impinged on the quality of the finished recording.

While their arguments may compel us to agree that there are significant differences between original and recording, we may not need to attribute the same importance to those differences. Altman's claim that "recorded sound creates an illusion of presence while constituting a new version of the sound that actually transpired" need not lead us to conclude that the "original" sound is of any great theoretical importance, since, as he argues, the recorded sound is effectively a "new version." Why, then, is the "original sound" so persistent? Williams's analysis of recording as a form of "hearing" ("we do not hear, we *are heard*") suggests an answer. At that moment, Williams (and Levin and Altman following him) effectively limits representation (and filmic narration) to the act of technological inscription and, in the process, analogizes it to a *simulation of human perception*.[16] Consequently, Williams imagines all sound representation to entail (and to be defined by) the *duplication* of a concrete act of real audition by an embodied perceiver present at the original event. That is, he assumes that the staging, performance, and presentation of the original are in no way a part of the representational process (it is "complete" before recording), and that representations always and in every case attempt to produce the effect of being present at those—necessarily real—events. Hence "representation" intervenes only after the fact as an act of pure "listening."

Williams, however, confuses things by introducing the original sound where it may have neither practical nor theoretical relevance. Since the sound in the movie theater is, by his own definition, *unique*, and its unique charac-

ter may be determined by variables as diverse as the original sound production, the acoustics of the original space, the microphone's positioning and
design, the quality of the recording equipment, the quality of the reproducing equipment, editing, and the acoustics of the space of reproduction (to
name only the most obvious), its specific sonic features are not determined in
any consistent, predictable, or rigorous manner by the so-called original.
That is, given a particular "original," it is impossible to predict what any copy
will sound like, and, vice versa, given a recording, there is no sure way to
reconstruct an original. As a result, the original may be of little theoretical or
practical use, just as actual profilmic conditions of image-recording (the
absence of a ceiling or a fourth wall, the presence of equipment and personnel beyond the edge of the frame, the incompleteness of an action, etc.) may
be irrelevant to the character of the finished representation. Edison showed
that the so-called original has only a functional importance, serving simply as
one step in the process of producing a satisfactory recording. No one was
ever meant to listen to Edison's "originals" nor were they *designed* for listening—they were designed to accomplish one stage of a multistage representational process.[17] If this is indeed the case there may be other, more pertinent
questions to ask of a particular sound recording that may speak to its specific qualities and its ideological significance in a more direct and salient fashion.
The key to the dilemma of "the original" and to its potential solution lies, I
believe, in the legacy of ideological critique represented by the work of
Theodor W. Adorno, who raised the question of the relation between original and copy quite cogently in his 1941 essay, "The Radio Symphony."[18]

ORIGINAL SOUND: RECORDING / REPRESENTATION

In this influential essay, Adorno sets out "to make a study of what radio transmission does musically to a musical structure," noting that "here would
appear the problem of the role played in traditional serious music by the
'original'—that is, the life performance one *actually experiences*, as compared
with mass reproduction on the radio."[19] He predicates his critique on a complex formal analysis of a specific musical form, a mode of listening deemed
appropriate to that form, and a series of inevitable technological transformations in order to argue for the absolute primacy of the original sound.[20]
However, he does not develop a universal or even more general sound theory based on rigid hierarchy of original and copy. Nor would such a theory be
without its problems, not the least of which is the problem of defining the

original itself with any rigor. While Adorno can point to the written score as the basis of *his* original, Williams, Altman, and Levin have demonstrated that different points of audition can divide even the apparently "single" original sound into a potentially limitless number of different and unequal events, and therefore even their original resists any purely physical definition.

In both "The Radio Symphony" and the related "Social Critique of Radio Music," Adorno skirts the dilemma of the "original" while still detailing the differences between listening to a symphony (specifically, a Beethoven symphony) in "life performance" and in one's home, via the radio, in order to insist that broadcast does not simply transform the symphony, but does so in such a manner as to corrupt its "essence." For Adorno, broadcast alters musical parameters to such an extent that the listener is barred from a performance's full realization of the score. Levin provocatively generalizes this analysis beyond the symphony and its objective structure (which, according to Adorno, "constitutes the concreteness of the musical phenomenon even more than the sound")[21] to encompass all sound and all applications of recording technology, whereby sounds are "put in an apparatus which spits [them] out in a digested, blunted, and conventionalized form."[22]

However, Adorno's essay discusses *only* the symphony (which he identifies as the most highly structured of all forms of musical practice) and how symphonic *structure* is altered by recording and broadcast. For Adorno, music is objectively definable by the specificity of its part/whole relationships, and their dynamic processes of structuration.[23] Hence, any acoustic transformation that affects the *perception* of that structure is, axiomatically, detrimental since it interferes with the mode of listening deemed appropriate to symphonies.[24] Adorno's account of the role played by dynamics in the symphony's production of the "unity of a manifold as well as the manifoldness of a unity," for instance, stresses the importance of *absolute* volume and *absolute* dynamic range to the developing structuration of the symphony. Robbed of both appropriate volume (scale) and range of intensities by the radio's insufficient electronics and the living room's acoustics, the symphony's unfolding realization is irrevocably damaged so that structured developments of motivic material are reduced to mere static repetitions of an atomized detail.[25] Such a reduction leads, inevitably, to a fetishistic and insufficient mode of listening.

It is important to note that Adorno criticizes neither the "technology itself" (and especially not that of sound *recording*) nor bemoans what it does to sound *in general*. One could easily conclude that for popular music there is *no* such essential transformation since "loss of tone colour and intensity, a reduction of the overtone series, a compression of the dynamic range, an overemphasis on melody, a de-emphasis on accompaniment, an exaggeration of contrasts, etc."[26] are *already* a part of popular music's structure. These attributes, and the form they serve, may in fact be enhanced by recording. Indeed, in "The Radio Symphony" Adorno suggests that jazz, chamber music, pre-Haydn symphonies, and romantic music might well be compatible with broadcast by virtue of their regressive structures, which "pre-transform" the music in much the same manner the radio does.[27] Adorno even argued, in 1969, that the LP had become the ideal form for opera presentation because it separated the music (essential) from distracting spectacle and anachronistic and nostalgic traditions of staging (inessential).[28]

Adorno avoids blaming "technology itself" for other reasons, too. Logically, he would have been just as concerned had a new fad in concert hall architecture or a particular manner of conducting caused similar transformations of the music.[29] Listening from the lobby of a hall accomplishes much of what he finds objectionable in the broadcast, and undoubtedly Adorno would find the lobby as poor a point of audition as the individual living room, although not entirely for the same reasons. The question, for Adorno, is not simply one of "original versus copy." A poorly presented original is no better than a copy if it similarly distorts the work. Moreover, if listening itself has "regressed" to the extent that it is now inadequate to the work, neither concert hall nor private parlor will solve the problem.[30] For Adorno, the issue is not technological mediation per se, but a compromised and degenerative mediation, whatever the source.

Adorno takes great pains to develop the powerfully interdependent nature of musical form and mode of listening, and insists on their specificity at every point. Can Adorno's analysis, therefore, be generalized to cover all recording? Adorno seems to answer no, since his own approach examines how broadcast technology intervenes in a very specific network of interlocking cultural practices that shapes the auditor's relationship to the presentation, perception, and evaluation of a particular musical form. These elements together determine which musical parameters are *pertinent* to this or that particular act of listening. To analyze another musical form, Adorno claims, would require a separate analysis of form, technological transformation, and appropriate mode of listening. Symphonic (or "structural") listening demands that the most subtle of some differences be audible and, in a sense, legible, yet this focused sensitivity cannot simply be extended to serve as a general model for how we deal with all sounds. Like other forms of audition, music-listening simultaneously emphasizes and *ignores* certain features of a sonic event in favor of others—indeed, the mode of attention and perception brought to bear on the experience produces its effective shape and formal hierarchy. In contrast to theories assuming that all perceptible differences between original and copy are always equally relevant, Adorno argues that different sonic forms call for different forms of listening. According to this view, an adequate theory must take into account the specificity of both the sonorous object and the appropriate mode of listening.

Nonidentity theory, however, assumes a single understanding of sound and a single model of hearing. While its model of hearing is sensitive to even minor acoustic variations, it consistently measures and evaluates recorded sound in terms of a theoretically prior and structurally autonomous "original" of absolute and irreducible sonic complexity. Recorded sound is therefore *not* "really" a unique event at all, but a parasitical repetition of a previous phenomenon. By defining sound recordings as partial, transformed, or to some degree *absent* with respect to the original, they present an almost Platonic theory of recording, where both truth and being decline as one moves toward the copy.

"The Radio Symphony" does not easily support such a move, however, for, as Levin points out, Adorno's "analysis is *not* a nostalgic 'fall' story, but a narrative of transformation."[31] According to Adorno's approach, there can be no valid reason for suggesting that one socially constructed practice of sound production and reception should ground the discussion and evaluation of *all* others.[32] Indeed, for someone like Metz who is more concerned with function and signification than with acoustics, there is no "loss" between original and copy because even the act of listening to the original sound is a directed one—directed toward identifying the source at the very least. To be sure, Metz offers a theory based upon one rather particular model of hearing, and one that is not necessarily going to render the most sensitive discriminations between sounds, but the nonidentity theorists assume their own equally biased model of listening—one that universalizes the acutely sensitive symphony listener. Since the presumed auditor is linked structurally to all other aspects of the representational process, this model entails a wide variety of repercussions.

Positing an unbridgeable gulf between original sound and copy, nonidentity theory is always and in every instance sensitive to the most minor acoustic variations, allowing us to make finely nuanced distinctions between any two sounds, recorded or not. It correspondingly tends to regard all such differences as inherently meaningful. Like the exquisitely sensitive symphony listener, who might truly feel his concert had been ruined because he chose to move his seat two rows in the wrong direction, the auditor implied by non-identity theory is—or should be—constantly aware of the absolute unique-ness of his experience, and its absolute difference from all other possible expe-riences. But, as Altman's discussion of different modes of auditory attention implies, such sensitivity is not characteristic of the way we engage with most sounds. When listening to speech, for example, we typically register differ-ences like "accent" and "emphasis," but we likewise *disregard* other differ-ences between different performances of the same words or sentences. While there are "material" differences between these performances, and it is surely significant that we ignore some differences but not others, these conventions of speech recognition are not necessarily of universal theoretical significance. They are functions of a mode of listening appropriate to a particular situa-tion, and need to be analyzed as such.

Our specific standards and practices of listening are always a function of particular contextual and often institutional demands and often appropriate only to those demands, hence no *single* context or set of evaluative criteria can—without elaborate justification—provide a reference point for theoriz-ing all others. By establishing one type of listening as the unquestioned stan-dard for all sonic events, nonidentity theory introduces an unacknowledged hierarchy of more and less appropriate forms of listening. Moreover, if rec-ognizing differences between a sound and its representation (or between two different instances of the "same" sound) were a matter of truly recognizing differences pure and simple, there would be no reason to order originals and representations hierarchically—each would constitute a unique sound event, and questions of "appropriate" forms of audition would be a separate issue.[33] Only by establishing the relevance of a particular set of evaluative criteria can we argue, as Adorno does, that particular technological transformations demand a hierarchical ordering of original and copy.[34]

It seems obvious to insist on a single model of hearing only if we simulta-neously theorize representation *as* hearing and if we define the technology as a surrogate ear. Williams, like Altman and Levin, predicates his critique of recording on its inherently "perspectival"[35] nature. That is, by dint of the devices involved, a recording can offer only one listening perspective on an original sound. A recorded sound, according to this account, has already "heard" the original for us, but its mediation has simply been hidden or for-gotten, and along with it, a kind of "preinterpretation." However, since any chosen point of audition on the orignal sound may effect transformations of just the same sort (this is the very basis of "perspectival" critiques), the defin-itive nature of microphonic or electrical mediation is questionable. We must consequently recognize that the space of the original, the volume of vibrating air—what we normally call a room's "acoustics"—*itself* "transforms" the sound, in the sense that it shapes it and does so in such a way that its work is inseparable from the sound "itself," and therefore inseparable from any subsequent mediations. Hence, any original event is already subject to "transformations" potentially identical in kind and in effect to those wrought by the sound recording and reproducing apparatus, and therefore the idea of recording technology "itself" becomes substantially more complex.

Moreover, these theories necessarily assert as axiomatic that all representations are records of literal perceptions and therefore are defined by an originary and more fully present listening experience than that afforded auditors in the movie theater. This approach simultaneously reduces the representational process to a mere act of technological inscription since it presumes (and in a manner of speaking *produces*) the fullness, autonomy, and purity of the "pro-phonographic" event. By insisting that the representation is (or strives to be) an act of pure, passive *listening*, the "technology" of sound representation effectively excludes everything prior to (and subsequent to) the moment of recording, and presumes that the original is therefore innocent of any role in the process—something Edison showed was impossible in practice. Only by equating recording with hearing can we construct a thing called the "technology itself," which simply takes our place at the original event and serves as a surrogate perceiver with whom we can identify. As Edison's case makes clear, however, "originals" destined for reproduction (say, the sounds on a movie set) may have only the slightest resemblance to the finished representation. This theoretical insistence on the autonomous original, therefore, needlessly privileges presence over mediation rather than simply illustrating transformation.

Although nonidentity theory provides us with a sophisticated method for discriminating between sounds, it entails a surprisingly limited understanding of technology. In the guise of alerting us to the fact that we have "forgotten" the mediation of the technological apparatus, nonidentity theorists themselves forget the mediations of the original—a mediation every bit as conventional as that wrought by the device. We might say, in fact, that before the appearance of a microphone, the symphony is *already* the product of an advanced sound technology—architecture—and therefore not innocent of technological mediation in its primary form.[36] Ultimately, the original sound can only stand as a theoretical absolute by dogmatically asserting that one socially sanctioned production of sound (the most ritualized), one experience of the sound (the "best seat in the house"), and one understanding of recording (simulated phenomenal presence) are somehow logically the essential nature of all sound representation.[37] Contrary to the claims of theory, which locates all the significant ideological work in what the device does to the original event, the primary ideological effect of sound recording might rather be in creating the *effect* that there is a single and fully present "original" independent of its representation at all.

If we were instead to expand our technological investigation to include the role of concert halls, for example, we would have to admit that the careful diminution or enhancement of various features of the sonic event (the very point of performing in specially designed spaces) constitutes a veritable technology of sound manipulation, as important as any microphone or loudspeaker. We would also have to agree, along with Adorno, that this manipulation is in no way extrinsic to the "sound itself." There is no pure original here; if we change the hall, we change the sound in its very essence. The problem of using the original, or "pro-phonographic," sound as a theoretical absolute is even more obvious in the case of present-day Hollywood film sound, where no one involved ever assumes that the finished sound track will attempt to duplicate the acoustics of the *set*. It is simply presumed that filmic sound space, like the image space, will be constructed.

Given that architecture shapes sound as thoroughly as the electronic portions of the recording apparatus, and that these transformations may even be identical in kind and degree, there is no "natural" or "logical" reason why one set of transformations should be accepted and the other demonized. Should we not also describe the process of performing pieces in a concert hall as "put[ting them] in an apparatus which spits [them] out in a digested, blunted, and conventionalized form?"[38] The hall surely "spits out" sounds in a highly conventional form (and for that we are usually grateful), but that the original space is not considered a part of the sound apparatus is perhaps the real blindspot here. Why we ignore this shaping is clear enough. The concert hall does not, conventionally speaking, "blunt" sounds but "enhances" them by presenting them in a form appropriate to a specific, sonic activity—listening to serious music. That these strategies of sonic manipulation escape "critical analysis," despite the fact that they are every bit as "transformative" as recording, is simply a function of their transparency *within the horizon of a particular social practice of sound and musical performance.* Thus we find ourselves, ironically, back with Metz at the movies, where taking a recorded gunshot for a gunshot is no more or less a problem than taking a theater flat for a wall, or a concert hall as the "normal" place to hear music.

THE TECHNOLOGY "ITSELF"

The question of technology, however, is not settled simply by noting that sound recording is not equivalent to the microphone. It also involves understanding a technology's place within larger ideological and historical patterns. One of the projects closest to apparatus theory's attempt to study "*the technology of sound itself*"[39] is Martin Heidegger's analysis of the technicity that he sees as characteristic of Western modernity.[40] His well-elaborated effort to think through the basic issues of such a project addressses many of our questions concerning technology and ultimately suggests that it may be useful to consider the cinematic apparatus as a part of that larger process he describes as objectifying the world through a structure of representation (*Vorstellung*). But what does he mean by technology? For all their emphasis on technology, Heidegger's texts do not easily support the conclusion that it is the *device*, the technical object *itself*, that frames the world in an objectifying manner. Although it seems correct that "irrespective of the particular content of the image, the photograph is a representation already interpreted, selected and ideologically 'framed' by its very technology,"[41] this does not mean that at every historical moment and in every historical use, the "very technology" is the same.[42] Heidegger himself suggests as much in his "Dialog on Language," where his Japanese interlocutor uses Heidegger's own example, *Rashomon* (1951), to claim that even in a non-Western film, "the Japanese world" has been "especially framed for photography."[43] However, to say that the world is framed *for photography* is not to say it is framed *by the camera*. The *device* can never fully embody the practices and institutions of "photography" nor can it ever be completely self-defining—it is always determined in part by larger structures of representation and knowledge.

Heidegger instead effectively argues that the entire process of staging a narrative with actors for a camera, destined for a mass audience, objectifies the Japanese world by making it subservient to the demands of an inappropriate representational structure. The whole project we might call "enframing representation" appropriates the world as an object for consumption in a form (the photograph, the commercial film) that is universalizing and trivializing in the same measure. As one symptom of the generalized technicity typical of modernity, photography "frames" reality as entirely commensurate to human understanding and as manipulable to human need. In the vocabulary of Heidegger's essay, "The Question Concerning Technology," it constitutes the world as "standing reserve"—as available to human use and wholly understandable in human terms. Technology therefore does not "itself" enframe objects because it too is something of an expression of the dominant historical configuration Heidegger refers to as *Ge-stell*, a positing of the world that frames it as available for human purposes. However, Heidegger argues that *any* relationship (not only those mediated by a technological instrument) that seeks such mastery of the world is indicative of technicity. According to this analysis, a specific historical configuration of *reason* dictates that the technology be deployed in a particular way.

Insofar as photography (or by extension, sound recording) produces an image that can be understood as "a systematic structure, a model or formation (*Gebild*) devised by man serv[ing] to explain that which it represents,"[44] it embodies the "frame-up" that Levin, for example, decries. Thus understood, "photography" refers not simply to a device but to an entire network of attitudes and practices determining the use of that device with the goal of producing images that render the world familiar and explicable through a structure of representation. When we invoke "sound recording" we imply just such a network of interrelated practices whose parameters are defined by standards such as "professionalism," and so on. Thus, when Adorno refers to a recording of a symphony, I can assume that such a recording was made with microphones directed at the orchestra rather than at the audience, from a distance rather than from the bell of the third trumpet, and in a relatively unobstructed space rather than from under a chair in the last row of the balcony.

Apparatus theory's implicit claim that there is an identifiable and stable technology producing uniform representational and ideological effect independent of and transcending any specific representational act now seems, I hope, somewhat reductive. Since there is no uniform and predictable set of transformations between every "original" and every "copy," there is simultaneously no "technology itself" that can be identified as producing any such uniform effects. Technological representation is a process that involves everything from the pro-phonographic staging of a sound to the final editing, mixing, rerecording, and playback. Any act of representation, even by means of technology, arranges its object in such a way as to make it "representable" according to a set of standards appropriate to an established and often institutionalized aesthetic practice. So, when recording music, for example, you record not just any performance but music presented in such a way that it can be represented "accurately" (to choose one standard). Music is "formed" by the hall-apparatus, and this forming is an essential first step in presenting the object ("music") *for (technological) representation*. Likewise, silence, soundproofing, and myriad other techniques ensure that on the movie set, the so-called original presents intelligible dialogue over all else. Every act of representation, by selecting only certain objects or objects in a certain form, or from a certain point of view, *pre-structures* its objects *for the device.*

Many of the theoretical problems discussed so far in this chapter disappear when, in a thoroughgoing way, we understand the process as sound *representation* rather than sound *reproduction.*[45] Although the recording of serious music tends toward the latter, it is not difficult to see that reproduction is simply a special case of representation, with no essential theoretical priority. In other words, we need to reconceptualize representation as an interlocking set of diverse practices and not simply as an act of pure inscription. Thus, we need to remain sensitive to the theories, standards, norms, practices, and implicit epistemologies of modes of representation employed by specific groups at particular historical moments. I propose, therefore, to look more carefully at the group of cultural professionals—sound engineers—who had the most power over the conceptualization, development, and implementation of sound recording and reproducing technologies within Hollywood filmmaking practice. Furthermore, these individuals, as part of a professional group with specific responsibilities to industry, were in a position to set both the technical and (because these were representational technologies) *aesthetic* standards of an entire industry. The integration of technicians—in fact, engineers—into the Hollywood studio system brought into conflict two well-established institutions (the phonography/telephony industry and the classical cinema) with opposed conceptions, traditions, and practices of representation. Although, at a distance, the conflict might not seem evident and the norms of classical filmmaking undisturbed, this was not at all the case. The peculiar history of Hollywood's transition to sound bears the traces of the clash between two different, but not wholly incompatible, modes of representation.

ORIGINAL / COPY: SOUND PRACTICE?

In the terms proposed by the sound theorists cited at the outset of this chapter, positions in the technician's debate can be polarized according to the general schema original versus copy or event versus structure. In fact, these twentieth-century models rename the simulation and inscription tropes of the nineteenth. Building upon a set of norms that were dedicated essentially to the ideal of perceptual simulation, transition-era debates adopted real-world norms to define their representational goals. Despite the fact that the majority of early Vitaphone films were based on the synchronous recording of *music* (and that Warner Bros. initially may have had no plans to make "all-talking" features), all sounds were ultimately recognized to be functionally subordinate to the *voice*.[46] That is, all sounds were treated on the model of the voice, ensuring that a pictured source "spoke" its characteristic sound. The most important aspects of a sound were assumed to be its origin, its identity, and its comprehensibility.[47] The essential task was to produce the effect of appropriate and recognizable sounds emanating from the objects on the screen, as if the sounds were, like the voice, a property of the depicted object.[48] In other words, Metz's concern with the identification of sound sources was, in effect, the same problem that concerned technicians during the transition period.

While the institutionally sanctioned goal of attaching recognizable sounds to sources may have served as an industrial mandate, the more general model of perceptual simulation that guided both basic research and representational practice continued to stress absolute, perceptual fidelity as the obviously paradigmatic objective. Soon, however, the Hollywood demand and the aesthetic principle would come into conflict. The simulation aesthetic seemed to call for a continuously receding, homogeneous, and "real" acoustic space centered on a clearly defined, single auditor. Effectively, it dictated that classical narration be understood as the manipulation of point of view (or "point of audition"), thereby literalizing a familiar heuristic for explaining the process. This model ran headlong into the competing demand for narrational clarity. In fact, many of the transition-era problems seem to have revolved around two competing and more or less incompatible models of sonic space and intelligibility—ones tied to different forms of sound practice and different representational norms, and which therefore supported different representational industries. Unlike their academic counterparts, but borrowing a page from Adorno, technicians from the phonograph industry elaborated a theory that encompassed an interconnected set of assumptions ranging from staging sounds, to technological inscription and reproduction, and, finally, to audition.

FIDELITY VERSUS INTELLIGIBILITY

The two general models of sound recording that dominated—and to a remarkable degree continue to dominate—technicians' ideas about sound representation could be called the "phonographic" (or "perceptual fidelity") model and the "telephonic" (or "intelligibility") models. Like its nineteenth-century predecessor, perceptual simulation, the former sets as its goal the perfectly faithful reproduction of a spatiotemporally specific musical performance (as if heard from the best seat in the house); the latter, like writing, intelligibility or legibility at the expense of material specificity, if necessary. A recording of an orchestra, for example, should try to preserve the reverberant space within which the sounds were produced, while a telephone, being designed with a very different social function in mind, would never be suitable for this. However, a "phonographic" recording could just as easily represent one of the "worst" seats, while still remaining "faithful" or specific. A telephone, in contrast, sacrifices acoustic specificity in favor of rendering speech clearly under widely varying conditions.

For each model articulated by sound engineers and researchers, the very conception of what a sound "is" differs in important ways. The "fidelity" approach assumes that all aspects of the sound event are inherently significant, including long or short reverberation times, ratios of direct to reflected sound, or even certain peculiarities of performance or space. The "telephonic" approach, not literally limited to telephones and voices, assumes that sound possesses an intrinsic hierarchy that renders some aspects essential and others not. As with our academic theorists, the relevant terms of comparison are uniqueness versus recognizability or event versus structure, and consequently, these terms presume different ideals of "good" representation.

Within the classical Hollywood style, we can find instances of both the "telephonic" and "phonographic" approaches to sound representation. According to the principle of narrative priority, dialogue recording tends almost uniformly from the early thirties on toward the telephonic, minimizing the amount of reverberation, background noise, and speech idiosyncrasy, while simultaneously maximizing the "directness" or "frontality" of recording, and the intelligibility of the dialogue.[49] Even when a speaker appears to turn away, a high level of direct sound often implies that he or she is still speaking "to" the auditor, because speech is understood not simply as an abstract sound but as a sound with a specific social and narrative function.[50] This statistically dominant form of recording differs markedly from the less common technique that has been dubbed point-of-audition (POA) sound, which tends to correspond more closely to the phonographic model of represented hearing.

In spite of its relative rarity as a technique, the phonographic or simulation model ruled the theoretical roost. This should not surprise us, since it lent itself easily to the reigning theoretical orthodoxies of the day. By insist-

ing on a physically real observer as the principle of representational coherence, it was easily understood as a variation on the "invisible witness" model
of narration. From Hugo Münsterberg and V. Pudovkin to André Bazin, classical film theorists have tried to account for narration by claiming that editing
(for example) mimicked the perceptions of an "invisible observer." In this
view, edits reflect the shifting attention of a real witness, and filmmaking
becomes the process of moving that witness around through space to the
most salient details of a scene. The frequently impossible spatial leaps and
changes of perspective characteristic of standard continuity editing give the
lie to this model, but it persisted nonetheless. As David Bordwell notes, this
theoretical expedient allowed filmmakers and critics to collapse the observer
with both the camera and the narrator. Furthermore, the ambiguity of the
term "point of view" (POV) allowed the conflation of both its optical and
intellectual senses. Thus, he argues, "the invisible-witness model became
classical film theory's all-purpose answer to problems involving space,
authorship, point of view, and narration."[51] Whatever inconsistencies (sometimes a real observer, sometimes an "ideal" one, sometimes an "omniscient"
one, etc.) or outright silliness the invisible observer produced, the model died
hard.

In addition to its conceptual and rhetorical benefits, the invisible observer's insistence on the foundational importance of POV responded to other
pressures. If we look back to the nineteenth and early twentieth centuries and
remember the extent to which the technical media disrupted customary
understandings of the relationship between perception and representation,
the model makes even more sense. While a filmmaker like Dziga Vertov
might celebrate the machine-vision produced by the camera and editing, and
extol its modernity and superhuman capacities, the experience of radical
shifts of space and time made possible by the cinema (and even the gentler
dislocations of continuity editing) presented the spectator with an unsettling
and profoundly inorganic experience of the visible world.

Even a conventionally edited scene might offer microscopic or bird's-eye
perspectives, and it would almost certainly jump back and forth across the
shoulders of couples engaged in intimate conversation. Even if not literally shocking, this was a novel and potentially confusing experience. By
assimilating these obviously inhuman processes to a thoroughly (if prob-
lematical) human model, critics, theorists, and classical filmmakers carved
a space for subjectivity in film narration and "anthropomorphized" its
more perplexing possibilities. In essence, POV "humanized" machine perception. In a larger sense, by focusing on character, adjusting framing to the

human body, and emphasizing psychological interiority and character-motivated POV, the classical cinema systematically worked to minimize the more disturbing tendencies of the new medium. It is therefore even less surprising that variations on this "anthropomorphism" would creep into technical discussions.

POA sound (like the POV shot) attempts to represent the experience of hearing within the diegesis, normally the hearing of a character. Sometimes this is indicated by muffling sounds, as if hearing them through a wall, by including a Doppler effect to indicate the rapid passage of a sound source (by an increase or decrease in volume indicating the approach or retreat of the source), or perhaps by giving a clear sense of the acoustics specific to a particular location. All of these different characteristics can be assimilated under Altman's term "spatial signature," which includes all those indicators of the spatial and temporal specificity of sound production and reception that characterize any recording as unique, and that create an effect of simulated perceptual presence.[52] Both the POV shot and POA sound represent spaces that are to be taken as diegetically "real," and as heard by an embodied perceiver within that space. It therefore implicitly confirms the classical cinema's emphasis on the human character, while simultaneously rendering its other historical norms more apparently "human." In a rather straightforward fashion, it simultaneously embodies the idea of representation as perceptual simulation and responds to the same pressures that elicited its appearance in the nineteenth century. The difference between a sonic space whose principal goal is the intelligibility of some sounds at the expense of others (foregrounding narratively important information against a reduced, generic background), and a space that is constructed in order to represent a particular real act of audition, embodies the basic difference between the telephonic and phonographic models.

In a typical Hollywood film, visual space, when not explicitly marked as a POV shot, is primarily a marker of the narrative or enunciative level of the presentation. In other words, its dimension, angle, framing, etc. are typically not understood as someone's actual vision (neither narrator nor character), but as an index of narrative emphasis. Unlike the POV shot, which is marked as a unique perception in a specific, diegetically real space, a typical shot can usually be replaced by another shot that represents a noticeably different view of the space. A typical shot's index of substitutability is much higher than that of a POV shot since it presents *narratively important* information rather than *perceptually specific* information. It is indicative, however, that sound technicians who call for perceptual fidelity in sound and image representation often call for an "invisible witness" model of cinematic narration as well. Such a heuristic seems obviously to encourage one to think of film-making as a stringing together of unique perceptions.[53] The sonic specificity implied by the standard of perceptual fidelity was quickly pressed into a carefully defined, and carefully delimited, role. It was soon harnessed almost exclusively to particular characters within a fictional world.

Sound recorded as if heard by a character—POA sound—became the primary instance within the classical system where the spatial characteristics of sound might manifest themselves (often, significantly, for narrative purposes). This however, was the exceptional case. In general, close frontal miking of actors, which minimizes reflected and indirect sound, became the norm for dialogue since, as one researcher put it, "In no case did an increase in reverberation cause an increase in articulation [i.e., intelligibility]."[54] Or as another phrased a related observation, "The quality of a recording is effectively independent of the reverberant characteristics of the set if the microphone is within approximately three feet of the speaker."[55]

Together, the two quotations indicate that in certain contexts a "good" recording could be associated with recognizable speech sounds, and dissociated from any strict sense of fidelity.[56] In fact, some researchers from the period began to argue that perceived "naturalness" seemed to be correlated with the presence of reflected sound, while "articulation" was associated with direct sound.[57] In other words, these articles suggest that indications of spatial specificity and naturalness were linked to certain forms of *unintelligibility*, or at least suggest the possibility that fidelity and intelligibility are not necessarily related. Although my argument here focuses heavily on speech, I would suggest that terms such as "intelligibility" have analogs in the case of sound effects as well, where something like "recognizability" or "identifiability" performs roughly the same function—that of comprehension.[58] A recording with a high degree of reflected sound, or some other indicator of spatial signature and temporal specificity, corresponds to an approach that considers sound an *event*, while closely miked sound, with a relatively "context-less" signature, corresponds to sound considered as an intelligible *structure*—as a signifying element within a larger system.

According to a 1931 article by Carl Dreher, there need not be any irresolvable conflict between the two approaches, however.

> Since the reproduction of sound is an artificial process, it is necessary to use artificial devices in order to obtain the most desirable effects. For exam-

ple, it is normal procedure to reproduce dialog at a level higher than the original performance. This may entail a compromise between intelligibility and strict fidelity.[59]

Clearly at stake are not the sounds "themselves," but as Alan Williams says, the signs of those sounds—even in the case of supposedly strictly accurate recordings. Nevertheless, the problem of strict fidelity arises again and again in other contexts. John L. Cass, for example, complains that the "illusion" is destroyed when a spectator becomes aware that despite changes in shot scale, the recordist has maintained "close-up" sound in order to ensure intelligibility, thereby violating the presumed norms of "sound perspective." Since "the resultant blend of sound may not be said to represent any given point of audition," the spectator is left feeling that he is experiencing the "sound which would be heard by a man with five or six very long ears, said ears extending in various directions."[60]

Here Cass is, of course, assuming that film narration works by the "invisible observer" method, stringing together a series of unique and spatiotemporally real perceptions. However, filmic narration, as Bordwell points out, does *not* need the supposition of a flesh-and-blood observer to perceive every element *for* us.[61] "Real" space, as it is perceived in actuality, is simply not an issue for most images on the screen: their scale and angle are functions of narrative emphasis, not of more or less precise perception. Cass's symptomatic insistence on the model of narration as perception, however, indicates the extent to which a model of precise perceptual duplication ruled the imaginative world of film professionals, helping to shape the practices of sound representation for decades to come, even if it was ultimately to be marginalized or rejected. A legacy of earlier encounters with technology, the illogical but persistent "invisible auditor" served to rehumanize the cinema when it most threatened to become inhuman. The resilience of this persuasive but misleading model also illustrates how reigning notions of filmic representation may come into conflict with institutional and industrial demands.

SPEAKING / SIGNING

The historical resolution to the conflict between fidelity and intelligibility returns us to the vexing problem of "the original," which functions as the center of both the technicians' and academics' debates, and specifically how these debates intersect with the role of the classical "functional equivalent."

The relationship assumed to obtain between the "original sound" and its representation is of no small theoretical or practical significance. Strongly held and generally unexamined assumptions about the identity of the technology in question, about its function, and about sound in general have profoundly shaped the way we, whether technicians or academics, frame our problems and our questions about representation. As should be clear by this point, both academics and technicians have tended historically to limit the act of representation to the act of inscription, and therefore have operated under the assumption that the technology "itself" was autonomous of the institutions, and even of the other practices within which it was enmeshed. Although these assumptions have allowed us to recognize the transformations wrought by any sound recording, and thus taught us to recast the process as representation rather than recording, all these theoretical forays have preserved an ultimately unsupportable emphasis on the pure and inviolate "original" against which we measure the degradations or alterations wrought by the technology "itself."

By positing such an original, one must also posit sound events that exist purely in and for themselves, full and complete, and likewise that all representation is a recording of wholly prior independent events which would exist regardless of the act of representation. However, since events on the screen attempt to present the fiction that there *was* an event independent of its recording, we risk simply reiterating Hollywood's effects in the guise of demystifying them.

By stressing the unique character of every sound event, its essential unrepeatability, we can better understand the phenomena of sound representation, and as nonidentity theory rightly points out, better specify exactly what material alterations occur in a *particular* representation of a sound. But this emphasis simultaneously encourages us to conceive of sound along the model of speech—as the effect of a singular act and intention, the reception of which is guided by the desire for absolute presence of the "original." What we want, it seems, is not simply to understand or to recognize this speech but to touch what is "behind" the sounds—presumably something which, unsullied by transformation into the common coin of language, is more real. Understood in this way, any recording, or any *writing* of this speech, inevitably disrupts its presence.

Despite recognizing a conflict between the naturalness and intelligibility of a sound recording, technicians argued repeatedly over the form of realism appropriate to sound representations. Such conflicts are symptoms of historically changing notions of representational reference—in fact, different understandings of the basic apparatus—and the resulting struggles had important effects on representational form, technical practice, and even devices. What is more, in the debates over what form of realism or reference representations *should* have, we indirectly witness the struggles of professionalization, and the emergence of a new discipline, as I will demonstrate in subsequent chapters.

Both Comolli and Altman have argued that new realisms can emerge only in relation to existing models of reality and of representation.[62] Therefore, specifying the relationship between a new form and relevant existing ones allows us to indicate historically how and why the new form emerged, while still conforming to established cultural categories. One can thereby explain the regulated emergence of certain forms of knowledge *about* the technology rather than the simple appearance of a radically new and self-defining form. On this view, the sound film would need to conform to certain relevant reality codes. Although technicians repeatedly made appeals to the spectator's body and to real perception, this is surely not the only, or even the primary, standard of cinematic realism. The theater, telephone, radio, and public address are all potential models with strong relationships to film sound, but perhaps the most important conventions for sound come from *silent* cinema. As I have shown in chapter 3, silent cinema had already established ways of dealing with problems of uniqueness versus intelligibility, with sound "space," and with sound source, and had already defined the roles "speech" should fulfill in a film.

Returning to the notion of textual function, it is useful to remember that Bordwell, Staiger, and Thompson argue that aesthetic and technical norms impinge on all Hollywood practice by shaping the paradigmatic substitutability of various techniques in any particular film. Techniques that perform the same, or nearly the same, function in a film are termed "functional equivalents."[63] With this argument in mind, we might ask what roles were preordained for sound even before its possibility as a synchronized effect.

As Bell technician Harvey Fletcher points out in 1929, from the very start even basic research into sound recording and reproduction was guided by a set of industrially determined standards. Although apparently researching "sound" in general, Fletcher makes it clear that even fundamental research worked under the supposition that "sound" was always understood as having a very specific role to play as a carrier of information.[64] Thus, the reproduction of speech was implicitly (and often explicitly) guided by the preordained role of speech as, above all, an *intelligible* conveyor of information. Although specific social *uses* of sound obviously must serve as a starting point for the development of any sound technology and its technical standards, the *particular* sound forms and uses chosen are nevertheless extremely important. Despite the wide range of possible roles sound had fulfilled historically, the one that ultimately dominated telephone research meshed well with Hollywood needs.

Technicians began their research into sound reproduction concerned with providing as accurate as possible duplications of completely prior events— essentially the model appropriate to recording and broadcasting classical music.[65] This approach fixed the position of the microphone as an audience member and allowed all the nuances of the space and performance to color the resulting recording. This model worked quite well for other forms of music, since the representational assumptions (fixed auditor, lack of hierarchy within and between sounds, recording as record of perception, etc.) did not conflict with the needs of this mode (although other representational models unsuitable for serious music might have worked just as well).

The absence of a practical system of amplification for the music and sound effects that accompanied silent films, and the tradition of romantic music, encouraged (or required) architects to design theaters whose acoustics were rich in reflected sound and had extended reverberation times. This undoubtedly added to the richness of the music, but as we all know, such spaces can render speech almost unintelligible. The problem is compounded if a reproduced *recording* of speech carries a strong spatial signature of its own. The addition or multiplication of signatures upon what we would normally think of as a single sound not only confuses the *sense* of the sound but, by marking the reproduced sound audibly with the spatiotemporal characteristics of the conditions of reproduction, the "uniqueness" of the original sound *production* is partially eclipsed by its *re*production, splitting the "origin" of the sound.[66] This same phenomenon was later to become the object of fierce debate among record collectors and audiophiles with the introduction of stereo recording in the 1950s. Opposing factions argued heatedly about whether musical recordings should capture the acoustics of the performance space (which would then be overlaid with the acoustics of one's living room during playback), or whether they should be captured "dry," allowing domestic acoustics to assume the role of the "original" space. Both camps argued that the other's aesthetic effectively "falsified" the origin of the sound, either by eliding the acoustics of performance or of playback.[67]

In the sound film, the problem of indicating a sound's source is largely solved by synchronization. However, Bordwell and Thompson, usually the most careful of scholars in matters such as these, list "fidelity" (of sound to source) as one of the basic categories of film sound.[68] Decades of tin-sheet thunder and coconut shell hooves prove, to the contrary, that fidelity to source is not a *property* of film sound, but an *effect* of synchronization. A gun firing on the screen accompanied by any brief, sudden, explosive sound *produces* the effect of source; it doesn't require it as a precondition. Every judgment of "fidelity," however, necessarily requires predicating a source for the sound (although even the "correct" sound, if out of synchronization, can destroy the very grounds of the fidelity judgment).

In the silent film, the situation was different. The primary intratextual way of representing sound, the intertitle, raised problems of sound's visible origins differently. An intertitle is affected neither by the space of production nor by the space of reproduction. As a form of *writing*, it is (almost) by definition *legible*. For any literate audience member, then, dialogue is 100 percent intelligible. However, as a form of writing, intertitles introduce another series of dilemmas. Primary among these is their relative separation from their sources. The voice, in the form of writing, is separated spatially and temporally from its putative origin—the moving lips on the screen. This formulation echoes one of the classical or Platonic ways of distinguishing writing from speech. Aside from the loss of certainty as to source, though, several other potential disjunctures may intervene. For example, intertitles cannot easily reproduce the "grain" of the voice, regional accents, differences in pronunciation, tone of voice, rhythms of speech, inflection, etc.—all the sorts of indicators we normally use to identify speech as a unique discursive and phenomenal event.

Mary Ann Doane sees the potential sundering of voice from body as the scandal that both silent and sound films must avoid.[69] To "make up for" the "lacks" of the intertitle, she claims, the actors adopt exaggerated gestures and facial grimaces, making the originally "audible" visible. In a similar fashion, changes in typeface, use of italics, quotation marks, exclamation points, and/or deviant syntax (all of which appear, for example, in *Wings* [1927] or *Old San Francisco* [1927], to name just two), as well as the bracketing of intertitles by shots of the same pair of moving lips, served to assign a source to an utterance and to make it as expressive of the character's emotional state as possible. In each case, the characteristics that mark the utterances as unique are indicated by making them visible, or more exactly, legible, and repeatable. As a corollary, those aspects of a sound that mark it as authentic are never simply self-evident "attributes" of that sound. Only their inscription within a system allowing or requiring them to become legible can give them a semiotic import within that system. Thus, the supposedly unique attributes of an original sound become significant and, to a certain degree, perceptible as such only by constituting them as *signs*—precisely that which is *not* unique.

The potential "scandal" of writing is therefore overcome by . . . more writing. What Jacques Derrida analyzes as writing's characteristic *iterability*, which marks writing as *repeatable*, as quotable or citable in an infinite number of contexts, and thus as non-self-identical,[70] is traditionally seen as disrupting the presence or plenitude of speech. In fact, it is the intertitle's susceptibility to being repeated without alteration in different contexts which is specifically used to mark the protagonist's speech to his several sexual conquests as inauthentic in *Don Juan* (1926), the first Vitaphone feature. As I have sketchily pointed out, however, it is writing (as legible mark) that "comes to the rescue" to ensure the uniqueness or unrepeatability of the speech event. As in the sound film, sound "fidelity" is an *effect* of inscription. Ultimately, the intertitle's status as writing is characteristic of sound recording in general and may extend to the "original" events as well—all of which may be understood as "inscribed" rather than "spoken."

Interestingly, more than one technician from the transition period suggested intertitles as the appropriate textual model against which to measure sound representation.[71] In at least two cases, writers argue that the kind of particularizing reading, or in this case "listening," appropriate to a painting or a serious concert is not appropriate to either the printed story or the Hollywood film. Each author points out that fancy fonts do nothing to enhance a story's readability, and overly fancy ones can be a distinct disadvantage if they impair legibility. Although neither goes quite this far, each recognizes the necessity of a strict foreground/background hierarchy in the recording of sound to ensure narrative legibility, just as black type against a white page ensures print legibility. The shift from one mode of representation, appropriate to particularizing (or event-centered) interpretations, to another based (loosely) on the printed page, corresponds to a redefinition of the "object" or phenomenon being represented. At some point the meaning of the printed text became distinct from its material form, allowing editions in any font to substitute exactly for one another, just as sound was to become primarily a signifying structure rather than a unique phenomenal event.

The task, then, is to illustrate how these silent film techniques weigh on early sound film practices, and then later on the recording process, or the inscription of traces in general. As I noted earlier, conflicts over representing sound space were settled de facto by the adoption of the standard of close miking and a certain "frontality" of address. This solution ran against the major trend in microphone research, however, which sought to allow more and more distant miking to simulate hearing more closely.[72] The type of sound recording that results from close miking is relatively contextless or spaceless—that is, it is a sound whose "quality" (to repeat a technician cited earlier) "is effectively independent of the reverberant characteristics of the set."[73] In other words, such sound is spatiotemporally *unspecific*, legible beyond the original context, a permanent record of an ephemeral event, and therefore in some sense more like writing than like speech.

Effectively, this makes of a specific speech act, or more generally, sound event, a species of quotation—a transcription removed from its original context. A sound thus recorded can be infinitely repeated throughout the film with different images regardless of the space within which the supposed origin is situated, just as the same title card could be reused in *Don Juan* to indicate the mechanical and empty nature of his words. Such a recording style, by stressing the recognizability and function of sounds outside of a specific context, falls on the side of writing rather than that of speech, suggesting that a further analysis of the speaking/writing opposition might shed some light on the phenomena of film sound.

Like the intertitle and the quotation, however, such sounds threaten to float free of the specific contexts within which they are expected to function. Rendering sound recordings more like writing produced certain practical benefits, but it created other practical dilemmas. Having made sounds intelligible and signifying, they needed to be "reconnected" to concrete (but now diegetic) situations. If we consider everyday situations where the uniqueness of the written utterance, or simply its source, is an issue, there are common ways in which we overcome the iterability of writing and give it the status not of a repetition but of an original event. How do we make writing like speech? By appending a signature. I am suggesting that it is not *simply* an accident that the markers of a sound's spatiotemporal uniqueness have been named its spatial signature. Your name may be written by anyone, typed, photocopied, repeated endlessly, and it will still refer to you. Yet although typing your name at the bottom of a contract will *identify* you, it does not indicate your intention to uphold the contract. The contract will never be enacted, will never "happen," so to speak. Only by attaching your inimitable signature can you indicate your intention and effectively carry out your act of will. In its purest sense, the signature is an embodiment of writing as a unique event—of writing as speech.

In the difference between the name and the signature we can read the difference between the aural object and the aural event, the sound as recognizable across contexts, as iterable, and the sound as event, as unrepeatable. But it is the uniqueness of the original sound, its historical happening, and this singular event's relation to its representation (its repetition) that needs to be analyzed. The signature, in order to function *as* a signature, requires the notion of a fully present intention. The idiosyncrasies of the mark are what stamp this form of writing as historical, as happening once and happening at a specific time. Yet no signature could function if it were not, as Derrida has pointed out vis-à-vis J. L. Austin's argument in *How to Do Things with Words*, something like a citation.[74] Your "unique" signature is only valid through its comparison with some (obviously repeatable) model—it functions as unique only through its being verified as a copy. Signatures, which undoubtedly "work" every day as embodiments of unique intentions, are nevertheless marked always and already by a certain absence, or in other words, the event of the unique signature is marked in advance by its repetition.

The "original" sound event (of which the recording is a "sample," a "reading," "a partial correspondence," etc.) functions in roughly the same manner. The nonidentity theories of sound recording discussed here characterize the recording as a degraded form of an original event whose fullness can never be captured. The recorded sound may be fully recognizable and possess a remarkable degree of fidelity, but it is always a repetition, and therefore manufactures a false "presence." It is as if, happening later than the original, the repetition has worn away a bit, weathered a little. Yet is it not through our awareness of this "wearing away," this nonoriginality, that we are able to imagine an original at all? Is it *ever* possible to be present at the original event, fully? Alan Williams, Rick Altman, and Thomas Levin all tell us from the outset that this is impossible by definition. In fact, even if a sound *is* never repeated, and therefore appears unique, is it not always (through the inevitable reference to the horizon of its recognizability *as* unique) marked by the possibility of repetition, by iterability, by a certain form of absence? Like an intertitle, or like POA sound, isn't the "originality" or the unique source of the sound simply an effect, constructed by the emergence or inscription of differential, signifying marks?

THE "ORIGINALITY" EFFECT

The character of this absence is what is finally at stake. It seems as if the nonpresence, the general iterability of the sound, is a negative characteristic, a barrier to uniqueness, or to fullness. At its most extreme, an approach emphasizing absence and repetition seems to deny that specific events "happen" at all, in a historical sense. This, however, is decidedly not the case. Although the repeatability of a signature seems to indicate that all signatures are somehow *lacking*, it is this "absence" that is the positive condition of their capacity to function. Even a signature that appears never to have been reproduced has the possibility of repetition inscribed in it. In essence, for something to be original, its repetition must be recognized as conforming in some way to the original—or the original must possess a certain ideality that is capable of remaining recognizable in various contexts. Ideality then, would *depend* on repetition; would be, in a sense, an *effect* of repetition rather than a precondition for repetition.[75] Understood in this way, the repetition, insofar as it is recognized *as* a repetition, "creates" the original. The nonuniqueness or nonpresence of the event marks both the possibility of and limitation of ideality.

At an "original" sound event, we all recognize that each auditor gets a slightly different sense of the sound depending on his or her location and the directedness of his or her hearing. But doesn't this recognition itself imply that there is no strictly definable "original" event? That every hearing is in some way absent? In the case of speech, for example, we ignore certain aspects of the event because our hearing is directed toward certain ends, namely, the recognition of speech sounds for the purpose of linguistic understanding. The directedness of perception is often institutionally and conventionally determined, hence the possibility of a telephone's being engineered to have a poor frequency response. The "wear and tear" on the original sound is compensated for by the payoff of a culturally and institutionally relevant form of comprehension. We do not think of the sound as absent because the meaning is present. And so it is for most sounds. To paraphrase Ernst Gombrich, there is no "innocent ear." There is never a fullness to perception that is somehow "lost" by focusing on a portion of the event, by using the event for certain purposes, or simply by perceiving with some particular goal, say understanding, in mind.

Since every sound is defined by the specifics of the time and space of its production and reception, it would be logically impossible to define what any sound "is" outside of a particular manifestation. There is no definition of a "violin 440 A" except as an abstraction from particular occurrences—none of which will be adequately covered by this abstraction. Likewise, the spaces and times of production, and the types of sounds produced there, are, to a large extent, defined by some particular social form of sound production and reception. It seems obvious that what we perceive in any given situation is determined by the "goals" implicit in that act of listening. What counts as a significant element changes as we change our frame. When listening to a friend tell a story, we emphasize the comprehension of words, but were we concerned with our friend's health we might pay more attention to the quality of her voice—is it strained, unusually hoarse, tired, and so on. Likewise, a recording of a symphony may very well be inadequate for appreciating the formal complexity to its fullest, but the same recording could function perfectly well if all that were required were the identification of, say, the flute's melody in the first movement, or the recognition of a musical "quotation" from another piece. The point in these cases, which seem to me fully "legitimate" modes of listening appropriate to *some* contexts, is recognition and association—not musical appreciation per se.

Nevertheless, the sense that there is something more primordial than the traces or recordings persists. There is nothing particularly striking about this. The historical happening of the sound event, its spatiotemporal specificity, always appears to escape our apprehension. Whether we are in an auditorium, or listening to a relatively contextless, closely miked recording, or to one that stresses the peculiarities of the room, the event, in its fullness, seems to escape. The academic, technical, and vernacular theories prompt us to ask, "What is the 'real' of which these recordings are the traces?"

Perhaps it is in the nonpresence of the repetition, of the wearing away of the original surfaces, the decaying remains, that we recognize the possibility of the event itself. The recognition of absence by which we classify representations *as* representations, recordings *as* recordings, is a positive condition of possibility rather than a fault, and this is what so much theory misperceives. If we start from the premise that all sonic events are constructed, this need not open us to the abyss of absolute indeterminacy or relativism. These absolutes are just as unattainable as absolute presence. We do, however, open up the possibility of explaining specific effects of "authenticity," or of "immediacy," precisely *as* effects of a general possibility of "writing," and we can relinquish misty assumptions of the ineffable, unattainable, but for some reason all-defining, original. We need not relinquish the original, the real, or the authentic, but we must recognize that these experiences and values, too, are products of historically defined and mediated conditions, and that their emergence, like the emergence of representations of those phenomena, follow certain rules. We may well ask, then, to what is attributed or denied these qualities of "originality" and "authenticity," and *in whose interests?* Representational reference is finally, as it is in the writing of history, a question of right and law, a question of morality and politics, and a question of social ethics.

What remains is to develop a theory of representation that does not insist on a loss of being between original and copy, a theory that does not seek to define the referential effects of each and every representation by comparison with some logically unattainable ideal, and a theory that, conversely, avoids falling into any naive sense of equivalence between "original" and represen- tation. Understanding recording as a form of writing lays the groundwork for such a theory. Paradoxically, it is to Stanley Cavell's quotation about photog- raphy that we return to find an ironic statement of this "grammatology," if you will. By repeating and paraphrasing Cavell out of context, perhaps we can create a new intention for his statement. "In the sound recording the orig- inal is still as present as it ever was"—which is to say—just as absent.

CHAPTER 5

STANDARDS AND PRACTICES
Aesthetic Norm and Technological Innovation in the American Cinema

One of the most important characteristics of the classical cinema is surely its stability over time. As I argued in chapter 3, scholars such as David Bordwell, Janet Staiger, and Kristin Thompson have shown how a paradigm of bounded and hierarchically ordered formal options structured filmmaking during the classical period, and how the dominance of a particular form of efficiently conveyed—and profitable—narrative ensured the paradigm's hegemony. They further show how the structure of industrial production ensured the interchangeability of personnel and therefore a kind of stability regarding the practical aspects of filmmaking.[1] While this may be an adequate account of the resulting formal and industrial paradigm, it does not explain how that paradigm came to be internalized by those workers whose everyday task it was to produce the representations on which Hollywood thrived. Although in the abstract it is easy to understand why the economic success of clearly told stories of goal-oriented, heterosexual, middle-class protagonists mandated the paradigm's primacy in a profit-seeking capitalist corporation, and that the formal conventions of what came to be known as classical continuity cinema were especially (if not uniquely) suited to this form, it is more difficult to explain how the representational assumptions and norms necessary to this style of filmmaking came to form the instinctive or obvious solutions to the various representational problems raised by any particular script.

Previous attempts to describe these formal conventions as, for example, a collection of arbitrary "codes" have more often than not led us away from the kinds of analyses that would explain how the classical paradigm actually came to regulate aesthetic practice on the set. Teach a group of students the rules or codes of continuity cinema, and they will be able to dissect any number of scenes from any number of Hollywood films. Chances are, however, they will not be able to *make* a film whose spatial constructions and match-on-action cuts have the fluidity and clarity of a Hollywood film. For technicians, it is not a question of *understanding* these films and their norms per se, but of internalizing them as a material form of practice—a job, if you will. This complex and flexible form of knowledge is crucial to workers whose livelihood requires them to make the "right" decisions about lighting, framing, miking, and editing from nine to five, six days a week, yet it is nearly impossible to formalize.

Nevertheless, the norms of classical continuity construction *were* internalized by workers in the studio system as what Pierre Bourdieu calls a "durably installed generative principle of regulated improvisations,"[2] or, more simply, as a consistent and coherent set of predispositions toward a particular kind of practice. Workers do not consciously devise these norms, these "ways of doing," nor do they exhibit a conscious mastery over them—they arise out of preexisting, objective, material conditions of the social world. As a result, such practices have an objective significance that outstrips the conscious intentions of the practitioners, yet the practices themselves cannot simply be reduced to those objective conditions. Instead, they exhibit a regularity and consistency, perhaps even a structure, that is *compatible* with the conditions that produced them. The dispositions toward practice characteristic of Hollywood technicians are, in Bourdieu's words,

> the universalizing mediation which causes an individual agent's practices, without explicit reason or signifying intent, to be nonetheless "sensible" and "reasonable." That part of practices which remains obscure in the eyes of their own producers is the aspect by which they are objectively adjusted to other practices and to the structures of which the principle of their production is itself a product.[3]

By citing Bourdieu's perspective on this issue, I do not mean to suggest that technicians do not consciously try to conform to the conventions of continuity or to standards of beauty, or that they do not work with an eye toward the significance of their decisions. On the contrary, I believe this consciousness, indeed self-consciousness, is an integral part of the modus operandi of the Hollywood professional. Further, much of the character of classical films can be explained as deriving from conscious attempts to adhere to these widely recognized norms. What I do mean to suggest, however, is that even these intentional choices are structured in advance by conditions of which the worker is necessarily unaware. The shooting angles selected by a cinematographer may be explained by an adherence to, for example, the 180-degree rule. However, the implicit values shaping the decision to shoot the faces under particular lighting conditions, the sense of the proper way to shoot a shot–reverse shot sequence, or most simply, the very bounds of the obvious, the reasonable, the possible—the taken-for-granted—are not consciously or intentionally decided. To be sure, they are all compatible with the articulated principles of filmic construction, but not fully explainable by them.

The analysis to follow will concentrate on a very concrete and particular moment within the history of the American cinema: Hollywood's transition to sound film production. The period from 1926 to 1934 saw the film industry undergo the most extensive transformations in technology, personnel, formal conventions, and mode of production in its history. As several scholars have pointed out, however, the classical continuity system emerged at the far end of this transformation relatively unscathed.[4] Nevertheless, I will argue that this apparent lack of change does not permit us to ignore the processes by which a new technology and a new set of representational standards were integrated into an existing industrial mode of production. The transition period is especially enlightening because the intersection of two major representational industries—Hollywood and the phonography/telephony businesses—allows us to highlight the otherwise occulted norms guiding representational practice in both fields. The clash between competing norms of representation as embodied in technicians' differing standards of correct practice forced each industry to examine the logic underlying the apparent obviousness of its own aesthetic ideals. The self-evidence of these norms comes under serious scrutiny as the two groups struggle to articulate their own rationale for conventions of practice and their justification for maintaining these conventions in the face of overwhelming pressures to change.

As each group attempts to clarify and legitimize the practical norms that define it professionally, we may examine the role that informal or vernacular theory—which we might also term aesthetics—plays in the regulation, justification, and reproduction of workplace practices. We may discover the extent to which these articulated goals and standards neither match nor fully account for the variety of practices they ostensibly authorize. In addition, we can see how ideas of professional identity are complexly imbricated with allegiance to a particular aesthetic and, I would argue further, the technician's "misrecognition" of himself in the ideals it represents. The technicians' mis-recognition is by no means a result of their ignorance, but rather of the historical relationship established between science, engineering, corporate capitalism, industrial forms of research and production, and labor relations. Therefore, I will make larger claims about the role played by aesthetics in science, in basic research, in the structuring of representational industries, and in the development, implementation, and normalization of representational technologies. Ultimately, it is not so much the devices that define the cultural and historical impact of a technology than the practices that regulate, define, and determine it that are of utmost importance.

THEORY AND PRACTICE

During Hollywood's transition to sound, technicians' duties often seem almost evenly split between working on the set and writing theoretical treatises on sound representation. Rarely have technicians been so forthcoming with their opinions on the logic and conceptual bases of filmic construction, and even more rarely has the theoretical arena seemed so central to Hollywood filmmaking. Page after page in scientific and industry journals emphatically promote competing aesthetic models based either on phonographic fidelity or telephonic intelligibility, but why? What function did the articulation of aesthetic norms and standards play? Far from being incidental or epiphenomenal, technicians of the period seem nearly obsessed with articulating their positions on questions of representational illusion, accuracy, propriety, and validity. Advocates of competing models of sound representation justify their nearly antithetical aesthetic allegiances in the name of the same putative standard—a supposedly transparent "realism"—despite the utter incompatibility of their different norms of recording and reproduction. Put more complexly, each naturalizes his own ideals of practice by demonstrating their compatibility with a *particular* notion of representation that is described as obvious *and* as scientific, and which comes to stand as the paradigm for *all* acts of representation, no matter how diverse. Realism of a very particular sort thus served both to structure technical and aesthetic debates and simultaneously (if circularly) to measure the validity of practices by masquerading as a universal category of evaluation.

The importance of realism as a category of analysis and evaluation was not restricted to the field of aesthetics, but infiltrated and shaped the course of industrial research and the development of techniques as well. Bordwell and Staiger, for example, have pointed out that "realism" was explicitly adopted as an industrial goal, but they add this proviso. "As for realism . . . this too was rationally adopted as an engineering aim—but wholly within the framework of *Hollywood's conception of 'realism.'* "[5] In fact, it is precisely because of a *conflict* between Hollywood realism (which stressed formal unity and narrative plausibility) and the (perceptual) realism advanced by engineers coming from the phonograph, radio, and telephone industries that the transition from silent to sound cinema is so complex and interesting. Despite their common recourse to the standard of realism, we might even go so far as to say that the dominant model of representation in each community was so at odds with the other that effective collaboration between them seemed almost ruled out from the start. However, the sound engineers' professional identity was so completely bound up with their notion of perfect representation that the compromises between them and their Hollywood counterparts necessary for an efficient system of sound film production required complex negotiations. In other words, workplace relations were worked out, in part, within the field of aesthetics.

The relationship established in this period between the theoretical and practical realms, and between the sorts of statements appropriate to each, is indicative of shifts in the technician's social, economic, and professional position. To demonstrate this, I will examine the category of realism so central to both the representational practices of the period and to current ideological critiques of Hollywood and will show it to be multiple in nature and subject to fierce debate. Rather than a unitary and stable category of bourgeois ideology that floats above all representational practice, determining it in a uniform and insidious way as apparatus theory suggests, the category of realism is one of the prime sites of cultural struggle and appropriation since it serves to legitimate representatonal regimes and reaffirm dominant understandings of the world.

The two understandings of realism in sound representation basic to the transition period embody different conceptions of the epistemological and referential properties of sound representations felt to be constitutive of "good" representational practice in general. Theory, which could easily be understood as secondary to the real relations and functioning of the social world, is precisely the terrain upon which certain terrifically important cultural and political struggles are fought—battles over the nature of acoustic and visual reality and, as I argued in chapter 2, over the proper relationship between the senses, technology, and representation. If nothing else, historical debates over realism set the boundaries for the manner in which a recorded sound could be understood to refer to the audible world, and therefore authorized a circumscribed range of "legitimate" understandings, uses, and practices.

The historical development of the American recording industries and the rise of a particular sort of engineer within these industries almost *required* the debate over proper representation to take center stage because the theories implicitly held by those engineers helped to structure the entire field of aesthetic goals and options. It also shaped the course of technological development in specific ways. Thus, the connection between theory and profession is far from arbitrary. Ultimately, the changing contour of the sound representation debate also indicates the changing nature of the engineer's perceived role. As engineers from nonfilm corporations came to perceive their own identity as tied to the corporate success of Hollywood studios rather than, say, personal achievements, or the success of Bell Labs, they became "sound men" rather than engineers.[6] Concurrent with, required by, and to some extent, constitutive of this shift is a shift in their standards and expectations for sound representations. By investigating the contradictions between initial theory and resulting practice we can, perhaps, reimagine the link between social structure, text, and subject posited by apparatus theory[7] without having to resort to the vague pressure of an ideological demand for realism. The link between social relations and representational norms is, I believe, far more material and demonstrable.

THE "INVISIBLE AUDITOR"

What exactly *was* the dominant representational model for sound? An early and representative statement of such a theory comes from Bell researcher Joseph P. Maxfield (soon to head Bell's ERPI [Electronic Research Products Incorporated] division) in a 1926 article on sound recording:

> Phonographic reproduction may be termed perfect when the components of the reproduced sound reaching the ears of the actual listener have the same relative intensity and phase relation as the sound reaching the ears *of an imaginary listener* to the original performance would have had.[8]

This could be called the "invisible auditor" approach to realistic sound representation. As late as 1934, and in spite of years of practical experience to the contrary, Harry Olson and Frank Massa of RKO claimed that the ideal record-

ing/reproducing situation would involve placing dummies with microphones for ears in different places around the set and recording multiple
tracks. Theatrical playback would require multiple stereo tracks, with each
signal routed to the seat in the audience corresponding to the literal position
of its respective dummy at the original performance on the set.[9] Beyond its
extraordinary technical complexity (and cost), this model implies several
rather startling beliefs. These writers assume, first, that movies work by offering discrete individuals something like a theatrical experience of each shot or
action; second, that the spectator identifies his or her "position" with that of
the mike/camera; and third, that movies are a succession of discrete,
autonomous, and perceptually specific "observations." Perhaps most significant, these engineers assume that the film spectator/auditor is literally a part
of the same space as the "original" performance. This model further implies
that the space and acoustic quality of the *set* are identical with those of the
represented world. In brief, the mechanical eyes and ears of film production
are linked in a surrogate and wholly simulated body located in identical posi-
tons on both the set and in the represented world. While using a different
vocabulary than their nineteenth-century counterparts, they clearly assume
that sound recording is a simulation of real human hearing, thereby familiar-
izing the inherently alienating process of sound representation.

One practical result of this aesthetic of absolute perceptual fidelity is that
each take is to be treated as a unique and autonomous event rather than a
part of a more or less homogeneous series. Hence, the acoustic quality of
every take was assumed to be a function of its unique conditions of production, and thus not necessarily related to the quality of the preceding or following take. The basic representational unit of coherence from this perspective was therefore the shot (understood as a discrete perceptual experience)
rather than the scene.

To put this yet another way, the specificity of each profilmic performance—the actual space, the actual distance of the camera/mike from the
actor, and so forth—provided the principle of representational unity for each
take. In practical terms, the invisible auditor or embodied-audience-member
approach entailed a variety of related concepts and techniques, including the
often-expressed desires for "scale matching" and for sound "perspective,"
which entailed "looking" and "listening" to the profilmic performance from
the same position. Although the practice of systematically matching long
shots with distant miking and close-ups with their acoustic counterparts was
probably never practiced with any regularity, much of the technical literature
continually stressed scale matching as an "obvious" goal, since it respected

the bodily integrity of the supposed "observer." Joseph Maxfield offers a suc-
cinct statement of the principle in a 1930 essay for the *Journal of the Society of
Motion Picture Engineers*. "The problem to be solved is that of obtaining a
sound record which correlates with the picture in such a manner that a member of the audience is given the illusion of being an actual spectator in the
scene."[10] Following Maxfield's lead, in 1930 John L. Cass, an RCA engineer,
decries the use of multiple microphones for a single shot since the resulting
sound record represents the sound as "heard by a man with five or six very
long ears."[11] Likewise, in 1931, Carl Dreher suggests that "good reproduction
requires a loudness level approximately equal to what a normal auditor
would expect on the basis of his experience."[12] Although reasonably referring
his standard to auditor expectations, Dreher neglects to point out against
which experience to measure the appropriate volume of, say, romantic conversations between couples we have never met.[13]

Of course, this particular version of the theoretically ubiquitous "invisible
observer," which assumed a literally situated relationship to both profilmic
performance and finished film was terribly ill-suited to narrative feature films.
Pairing the mike and camera in close-up gave great prominence to the dialogue, but a cut to a long shot would introduce an enormous amount of
reverberation while simultaneously risking a loss of dialogue intelligibility.
The sudden changes in sound texture created by this approach were clearly
inappropriate to the more homogeneous character of classical representation
and narration that assumed compromises between literal fidelity and narrative clarity such as instituting narratively determined spatial hierarchies
between the sharply focused foreground and the comparatively indistinct
background. Indeed, as early as 1929 mike booms had been designed to follow actors around the set, negating in practice the theoretically dominant
invisible auditor principle.[14] The use of directional mikes and soundproofing
also put the lie to the model of utterly faithful sound duplication. So why did
this model persist well into the 1930s? What did it offer?

Perhaps the best answer is that scale matching offered a coherent set of
tools and predispositions for conceptualizing, describing, ordering, and eval-
uating representational practices. So, for example, when attempting to
explain the concept, Wesley C. Miller of MGM tells his readers that

> the amplification of the sound must be just enough to fit the picture size.
> Probably the best way to express this is to say that any combination of pic-
> ture and sound must be so proportioned that the latter sounds natural
> coming from the artificial person on the screen.[15]

By offering the description of an effect in the guise of its cause, Miller expresses a desire shared by dozens of other engineers, namely, to describe, in terms of realism or illusion, the effects of classical narration. That is, rather than recogize that the governing criterion in a given representational situation was narrational, and the goal of a particular technique rhetorical, engineers routinely appealed to the necessarily vague but presumably objective aesthetic of "realism." Given a particular representational problem for which his aesthetic is unsuited—say, the incompatibility of the phonographic model of sound space with the demands of classical image editing—Miller reconceives the necessary representational modifications through the intellectual, perceptual, and cognitive categories readily available to him. Although another set of categories would not have been beyond his grasp, those that were ready-to-hand, and ingrained as habits and basic dispositions, were far more comfortable and familiar because based upon standards appropriate to phonographic recording and listening. Thus, the invisible auditor model of cinematic narration offered a convenient logic of day-to-day practice that was consistent, logical, practical, and simple, and which ordered the almost infinite possibilities of sound representation into a small group of probable and "good" techniques while excluding the rest as improbable or simply bad. Its universal application allowed theoretical explanations for qualitative judgments—it gave aesthetic evaluations a quasi-scientific basis. Simply put, it offered mental equipment, mental habits, and a standard for the profession.

AESTHETICS AND "BASIC RESEARCH"

Now any number of norms or aesthetics could have performed this function. Why the perceptual fidelity model to the exclusion of others? In retrospect, this "mistake" seems tightly bound to the engineer's status in institutions like Bell Labs and, later, in the studios. One of the standard explanations for the "delay" in integrating recorded sound with the classical cinema has been the belief that sound engineers insisted on dominating all decisions on movie sets to the detriment of, for example, familiar acting and shooting techniques. So intense was the conflict between sound engineers and cinematographers that both groups referred to it in military terms and both vociferously blamed the other for cost and schedule overruns. The legendary conflicts between sound engineers and virtually every other worker on the set bear witness to struggles over professional identity and responsibility and over the representational standards developed over time within the competing technical and industrial cultures. The very fabric of these professions is woven out of the aesthetic norms and theories held by its members and implicitly respected in their day-to-day practice, and these norms shaped not only the evaluation of devices and techniques but also helped set the agendas for corporate investment and research.

The representational norms characteristic of engineers within the sound industries were partially, but definitively, determined by the economic situation of the larger corporation. By setting the objective conditions for research, development, and implementation, these corporations determined in a practical sense the limits of the reasonable, the thinkable, and the possible for its employees. Even though the research agendas of sound researchers at other institutions might, in fact, have differed little from those at, say, Bell Labs, the latter company wielded enormous power in setting these agendas. Bell researchers did not *skew* their research to suit some arbitrary needs of their employers, but, to quote Bourdieu again,

> because the dispositions durably inculcated by objective conditions . . .
> engender aspirations and practices objectively compatible with those
> objective requirements, the most improbable practices are excluded . . . as
> *unthinkable*, or at the cost of a *double negation* which inclines agents to make
> a virtue of a necessity, that is, to refuse what is anyway refused and to love
> the inevitable.[16]

The bulk of research and commercial development in the sound industry was carried out in firms like Bell Labs, where economic interest in all phases of sound technology, from mikes to amplifiers, to disks, to loudspeakers, encouraged certain forms of commercial exploitation while it downplayed other options. Particularly profitable forms were the Orthophonic Victrola and the Vitaphone disk-based system for sound film production, both outgrowths of Western Electric's long-standing interests in the music industry. Given that early research into both recording and reproducing was geared primarily toward music, the heuristic of the invisible auditor was a fairly obvious one to adopt. It made perfect sense in phonography to duplicate as closely as possible the experience of an auditor in a concert hall, sensitive to all nuances of tone, performance, reverberation, and volume. So important a criterion was the concert model of audition that advances in recording technology were often praised for allowing the recording situation to approximate more closely the concert situation in physical and acoustic terms.[17] This standard meant that the represented event, acoustic space, timbre, etc. should duplicate the original, and that the auditor was conceptually and practically a part of the space of representation—literally a witness to the performance. Thus the demands of one social practice of sound production and reception, those typical of serious concert music, shaped the theorization of a whole range of sonic phenomena.

Adopting the concert-listener as the universal standard had quite a number of practical implications, ranging from microphone selection and placement to concerns about the propriety of sound editing and mixing (which were, as a rule, forbidden). Yet from the outset engineers seemed to ignore the letter of these implied laws in the face of specific practical exigencies. For example, Maxfield notes that the reverberation long considered appropriate for piano performance seemed excessive when recorded, and therefore suggests that some manipulations of the original space of performance "can simulate to a considerable degree *in the reproduced music*, the *effective* space relationships of the original."[18] Two significant contingencies are introduced here through the back door, as it were. First, we are forced to recognize immediately that we are dealing with a highly conventionalized and highly constructed original event. We might say that this is sound produced *for representation*—that is, with a certain representational, rather than "pro-phonographic,"[19] effect in mind. In other words, as we saw in chapters 2 and 4, the original sound around which the theory of "correct" representation is built is itself manipulated in order to ensure a certain *representational* effect. It is therefore in no simple way a pure original that can be either faithfully reproduced or "distorted" by recording. Second, it is the "effective space relationships" that are now preserved rather than actual ones. The latter point implicitly acknowledges that certain representational standards of accuracy or correctness can be derived only by reference to relations established within a given representation, rather than by way of some actual original performance. It further suggests that representational effects are not necessarily a function of that original, and that they are conventional, signifying relationships—not absolute ones. Thus no absolutely transcendental categories of evaluation determine effective fidelity, even among the proponents of this model.

Of course, in spite of its profits from phonography and film, Bell's primary investment lay in telephony, which obviously had acoustic requirements different from those of music recording. There a different practice ensued. Implicitly recognizing the social *use* of sound embodied in telephony, engineers felt entirely comfortable sacrificing 60 percent of the voice's acoustic energy (the lower frequencies) because they lost only 2 percent intelligibility in the bargain. The functional primacy of intelligible speech enabled telephone systems to reduce drastically the amount of power required for transmission while retaining the ability to transmit voices with acceptable clarity. Nevertheless, the sort of research that allowed such telephone compromises—basic inquiries into the nature of sound and hearing in the abstract, as they had been defined by science—paradoxically encouraged engineers and technicians to conceive of *all* forms of sonic representation, regardless of function, as involving actual, attentive listeners, uniformly sensitive to all measurable aspects of the sound event, and situated characteristically as audience members.

In other words, sound engineers assumed that all sound representations must take into account the physical and physiological circumstances of an ideally defined listening situation—more or less that of the educated symphony listener—and attempt to replicate those conditions in all their complexity (binaurality, perception of source position through reverberation, phase and volume differences between ears positioned a fixed distance apart on an unchanging head, an so on—that is, the scientific definition of the relevant parameters) in order to achieve a satisfactory simulation of actual presence at the original production of sound. Engineers uniformly insisted on the universal applicability of this fully embodied and fully attentive model of hearing, despite the fact that millions of people every day were listening to telephones with one ear, on instruments with severely restricted frequency characteristics, with no complaints. In the face of the telephone's remarkable integration into everyday aural experience, and its ascendancy as the dominant means of electronic communication, even the modifications to frequency response that saved transmission power and gave phones their characteristic tone was constrained, in one researcher's words, by "a loss of naturalness."[20] The lab that could measure speech intelligibility in percentages was restrained by the very unscientific concept of "naturalness," which not even they could quantify.

The use of so fuzzy a term in a highly technical engineering journal indicates not so much a philosophical or scientific failure on anyone's part, but rather the extent to which implicit and explicit standards of representational accuracy, legitimacy, or simply "goodness" guide even the most scientific of attempts at objective evaluation. Technicians' ingrained and relatively unexamined beliefs about what constitutes a good representation or the acceptable limits of deviation from that ideal profoundly shape both the character of the representations they create and the representational devices they design. As the case of the telephone makes clear (although much the same could be said of the microphone, the amplifier, and the loudspeaker), the very design and application of devices is measured against the sort of norms deemed "obvious" by the dominant representational model, however inappropiate it might be in practical terms.

The competing models of representation implied by phonography and telephony could or perhaps should have suggested that representational norms were based in part on a sound's social or cultural function, but telephony was nevertheless routinely considered a "special case" deviation from the ideal of absolute perceptual fidelity. So, rather than offering a competing paradigm assuming the primacy of communication or information, the telephone was subsumed under the fidelity aesthetic, with the result that this norm went effectively unchallenged since it seemed to apply even in cases where fidelity was not obviously a concern. Other cracks, however, began to appear in the phonographic or invisible auditor model, as when Maxfield counseled engineers to alter the characteristics of the original space in order to obtain an acceptable record. That is, Maxfield suggested they manipulate the original or pro-phonographic space in order to produce the *effect* of listening in the "proper" (or functionally appropriate) space. Here an adequate *representational* space is possible only by disguising the characteristics of the *actual* space, and the supposedly necessary link between actual and represented space is decisively severed. In fact, all around Hollywood carpenters were busy deadening sound stages to eliminate extraneous but altogether natural sounds that a fidelity-based model would, ideally, wish to preserve. These deviations from the ideal were not felt to be impositions on proper technique despite being acoustic manipulations as overt as multiple miking. I would suggest that by restricting their definition of *representation* to the act of *inscription*, engineers could comfortably maintain a certain adherence to the fidelity model while still allowing manipulations to the set and therefore to the final product. They were, after all, faithfully recording the resulting sonic event. Thus, Maxfield could continue his insistence on a fidelity model long into the 1950s when stereo musical recordings emerged.[21]

A variety of justifications of this sort helped to make the invisible auditor norm a bit more flexible in practice, though never really challenging it in theory. Since professional representational standards assumed that the engineer was an operator of equipment, or simply a recordist, the various processes of representation could easily be reduced in theory to the simple act of technological inscription. Although the auditor model assumes no prioritizing activity on the part of the recordist, the sonic hierarchy characteristic of an orchestral performance in a concert hall offered the recordist an event that was already strictly ordered so that a simple recording sufficed to achieve the desired hierarchical or rhetorical effect. A soundproofed set completely lacking in reverberation achieved a similar ordering of the sonic realm in the movies. Thus, an engineer could simply delimit his own field of responsibility to the act of recording, thereby assigning professional jurisdiction over the set to carpenters and designers (while nevertheless exerting decisive control over their work), and still adhere to his own ideal of proper practice. While this approach allowed engineers to achieve Hollywood's characteristic sonic hierarchy without altering their presumed standards, it did not result in the most flexible or efficient system possible. That system would await a redefinition of the relation between the profilmic event and the represented event, and of the basic level of the representational unity. Crudely put, technicians needed to escape the equation of unique profilmic event and unique representation and recognize, as Edison employees had two decades earlier, that the represented world need not correspond to any actual world, nor need the individual shot remain the basic representational unit. In the classical film the *diegetic* unity of a multishot ensemble typically takes precedence over the literal fidelity of any single shot. In short, engineers were forced to redefine their function as *representation* or construction rather than duplication.

AESTHETIC NORM AND PROFESSIONAL IDENTITY

When engineers arrived in Hollywood, however, a steadfast adherence to the norms they brought from phonography and radio quickly placed them in conflict with other technicians, especially cinematographers and editors. Frank Lawrence, then president of the editor's union, described this "war" as a kind of "Armageddon" caused by engineers, or "sound experts" who were "hopelessly ignorant of the existing public demands and high artistic standards of the motion picture production world." He amplifies his claim thus:

> Every blooming sound expert in the entire world is at present convinced that the only way to make satisfying sound pictures is to sacrifice every other feature of value in Filmland to the proper recording of sound. . . . [The] more experienced divisions of motion picture production [should not] permit such rot. . . . Sound experts will have to get in step with the motion picture fraternity.[22]

In other words, while such practitioners might legitimately be thought of as experts in their field, they had not yet realized that their field had changed, nor that the norms appropriate to one representational culture might be inappropriate to another with its own equally coherent set of standards and assumptions. Although sound engineers considered their own standards to be objective and universal because those standards were scientific, they neglected the extent to which even their scientific categories were shot through with aesthetic assumptions and were therefore anything but transcendental and neutral. Sound engineers' strict adherence to their self-imposed and undoubtedly high standards confirmed them as experts of a certain type, but it hobbled them when they moved to a new sonic culture, such as that defined by Hollywood's demands.

Almost in response to Lawrence, L. E. Clark addresses the conflict over aesthetics and technique explicitly in terms of professional standing. After noting that sound recordists were typically denied screen credit and that this low esteem enforced a cycle of degenerative neglect, he locates the root of the dilemma in the differing technical "cultures" to which cinematographers and engineers belong:

> Very few of us were originally from the studios; we came from the electrical laboratories, from the telephone companies, from radio broadcasting studios and chains, and from engineering colleges. We knew little or nothing of the conditions within the motion picture industry—and cared little about them. We were engineers—and proud of it. Our business was to install and operate the recording equipment, not to make pictures per se. . . . The sound man, [simply] places his microphone, and adjusts his circuits to get a good, commercial record—and lets it go at that.[23]

He argues in conclusion that such soundmen exhibit an "apathetic attitude toward the artistic phases of picture making."[24] Another possible explanation for the slowness with which sound engineers integrated themselves into the Hollywood system avoids assigning blame or labeling one group or another as less than professional, while it still accords with the evidence. Perhaps sound technicians' stubborn maintenance of engineering standards can be better understood as a strategy for protecting the quality of their work and thus their professional standing. Such apparent intransigence might then seem instead a refusal to perform their work in a less than professional manner. Indeed, there is a wealth of evidence to suggest that they felt themselves professionally under siege.

A cartoon in the August 1929 edition of *American Cinematographer* entitled "The Age of Alibi," humorously portrays the perception shared by all technicians that the new combination of sound and image was causing movie quality to suffer and that someone would be required to play the scapegoat.[25] Professional standing and even careers were clearly felt to be at stake over the issue as was, not incidentally, the continued profitability of the industry. The cartoon's second panel shows a sound man next to a cactus proclaiming, "AT LAST—the ideal place for my microphone," referring to the results of a contemporary survey conducted by the American Society of Cinematographers (ASC) that asked both cinematographers and sound engineers for their opinions on the current dilemma. When asked for the ideal microphone location for recording, sound technicians replied, "the middle of the Mojave Desert, unhampered by cameras, walls or any other disturbing elements."[26] (See figure 5.1.) Beyond the obvious renunciation of any recognizable standard of realism or illusion, the wish for an utterly

FIGURE. 5.1

The Vitaphone camera in its soundproofed "ice box." (© Warner Bros., a division of Time Warner Entertainment Company, L. P. All Rights Reserved)

unimpeded arena for sound recording indicates that the attempt to pair sound and image had placed the two sets of technicians in sharp and apparently irreconcilable conflict with one another.

In response to the discord, each group struggled to maintain its self-defined standards of quality by taking every opportunity to shape the conditions of the workplace and the techniques practiced there, and thereby ensure a high-quality (visual or sonic) record. In order to avoid further conflict and loss of profits, both the ASC and the Academy of Motion Picture Arts and Sciences (AMPAS) created forums for debate and reconciliation. AMPAS went so far as to set up a special Producers-Technicians Committee headed by producer Irving Thalberg and engineer Nugent H. Slaughter. In order to ensure that everyone on the set worked from roughly the same set of assumptions about sound recording, AMPAS also established a series of courses on principles of sound recording and reproduction that were open to personnel from all parts of the industry, from writers to cinematographers. Working through the Continuing Education Program at the University of Southern California, at least nine hundred employees completed the program in approximately two years. One of the chief merits of the program as far as academy executive Lester Cowan (and, the evidence suggests, Thalberg) was concerned was the opportunity to train current studio employees to become soundmen instead of importing them from other industries.[27]

Such classes and forums for discussion did much to alleviate tensions and delays on the set, but given the entrenched norms and practices of the studios, it was ultimately the engineers rather than actors, directors, and cinematographers who would have to compromise the most. While the engineer's supposed "intransigence" can be understood sympathetically as a resistance to Hollywood's particular form of mass production that maintained "craft" standards of work in a factory environment, in the end the engineer had to capitulate and, as it were, make a virtue of necessity. The phonographic assumptions shared by so many engineers were based on an inappropriately literal notion of realistic duplication, and thus could not serve as a model for sound representation in general or for film sound specifically. The processes by which engineers negotiated the adjustment of their own standards to the reigning model is thus important for understanding the power of the classical paradigm. Among other things, it indicates how the paradigm functions not only in a positive fashion by encouraging certain norms, but also how it necessarily effects and maintains a certain exclusion and thus ensures its own regularity.

NEGOTIATION, RECONCILIATION, REDEFINITION

In a 1929 article, Carl Dreher, one of the sound field's elder statesmen, points out that unlike actors and directors in the field of broadcasting, their motion picture counterparts had not yet adopted the "broadcast viewpoint." Radio personnel, he recounts, after similarly taking the demands of the sound engineer as a kind of professional affront (they, too, had professional performance standards to protect) gradually learned to "modify their execution for the sake of the microphone."[28] Taking a page from Edison's book, performers adjusted their techniques to the standards of the sound engineer, thereby ensuring that the representation, if not the (unheard) "original," was up to snuff.

In Hollywood, however, the situation was apparently quite different. Unlike radio, motion pictures do not rely solely on sound technologies for their existence, and therefore the engineer could not automatically assume the most powerful and authoritative role on the set. Rather than suggest that the various professionals on the set each adapt to each other's needs, Dreher argues that sound engineers capitulated early and too willingly allowed themselves to be forced again and again into impossible microphone placements in deference to camera technique. The resulting sound quality was poor and became a source of criticism directed at the particular sound engineer involved and, by extension, at sound engineers in general. As a result, the status of soundmen in Hollywood was much lower than it should have been, which hampered their ability to negotiate with directors and cinematographers.

Bolstering his arguments, Dreher points to the widespread tendency to blame sound technicians for all delays and inefficiencies on the set, regardless of whether they had actually caused them. To avoid such unwarranted criticism, he suggests careful note-keeping in order later to defend oneself and one's professional standing.[29] In other words, conflicting professional and representational standards were perceived as causing such extensive disruptions that the groups of workers with less institutional power were required to expend a great deal of energy simply avoiding a loss of prestige and workplace autonomy.

The various contemporary technical journals consistently point to the connection between these struggles over professional identity and responsibility and the respective representational standards developed over time within the competing technical cultures. To suggest that "theory" is in part constitutive of the technician's identity is not, however, to suggest that technicians necessarily engaged in explicitly theoretical debate over the nature of representation in general. It does suggest that something recognizable as theory shaped the standards and practices by which groups and individuals evaluated the worth of devices, techniques, research programs, and individual representations. From this perspective, it becomes clear that Hollywood conflicts between art and science or between cinematographers or "motion picture men" and sound "experts" or "mere technicians" were in fact complex processes of give-and-take and jurisdictional renegotiation between professional groups and individuals. Each was unwilling to compromise his own standards unless convinced by a compelling argument or result—and then only if this new standard could be justified as a "practical" or "artistic" deviation.[30] If Dreher's conflict were to be resolved, opposing groups would need jointly to retheorize their basic representational goals of the Hollywood film.

The belief that a single ideal might serve as a universal standard for all forms of representational practice is not restricted to technicians by any means. Bordwell, for example, argues that transition-era technicians in pursuit of "sound perspective" would try to match the acoustic qualities of sound to the scale of the image. "Engineers debated how to convey 'natural' sound while granting that strictly realistic sound recording was unsuitable."[31] As a result, Bordwell continues, engineers compromised by moving mikes to follow dialogue, dubbed sound after principle recording, and so on. As evidently correct as this may be, there is a rather peculiar assumption at play here. What, after all, does "strictly realistic sound recording" mean? Do "realistic" recordings require that the mike be placed close to the sound sources or at a considerable distance? Is "realistic" microphone distance determined by an absolute measurement or does one assume a real auditor whose presumed distance from the event guides microphone practice? Surely there can be no absolute standard of realism from which to measure deviations since there are an infinite number of situations under which a sound may be produced and/or received. Moreover, since there is no absolute connection between the space in which the sound is produced (i.e., the set) and the space represented on film, the type of sound "appropriate" to the latter might be wholly different from the sound appropriate to the former.

A growing number of Hollywood technicians began to acknowledge the impossibility of employing a single standard—"realism"— for all sound work, noting as early as 1929 that one should adjust his recording style to the type of performance being represented. For example, Wesley C. Miller argues that a vaudeville performer could be recorded more "intimately" and therefore louder than another style of performer "because of the nature of his act."[32] Eavesdropping on a quiet conversation, however, better suits the demands and expectations of the movie audience. Whether these specific examples are "correct" is less important than the implied recognition that in a practical sense no single standard should govern all recording situations. Moreover, by including audience expectations in his equation of representational correctness, Miller notes that different modes of address shape both performance style and recording procedure—that is, different sounds have different rhetorical functions.

Less than a year later, T. E. Shea argues that "[sound] requires not only much new apparatus, but new talent and technical training, new care and habits." Far from being a neutral site whose acoustic particularities must be respected, he adds, the "original" space should rather be understood as the first acoustic device for manipulating the recorded sound whose alteration "depends on the sound and scene to be recorded."[33] Another commentator writing in 1933 asserts that

> the importance of having the dialogue always clearly understandable goes without saying. Great care must be exercised at all times to have the sound effects recorded with the proper level to make the finished picture as realistic as possible. In an effort to *create realism*, we have used as many as sixteen separate sound tracks, each are carefully controlled as to level, perspective and quality, to make a pleasing composite track.[34]

Although still measuring his success against a largely undefined concept of "realism," Shea's realism is *created* and therefore attributable to the representation and not to the performance or to the act of recording. Also significant is his belief that the primacy of dialogue intelligibility now "goes without saying." Although from our historical vantage point this claim seems almost self-evident, the standard of realism most frequently articulated by technicians—the invisible auditor—assumed that if an actor "really" moved away from the mike and was "really" unintelligible, a realistic recording would duly record that fact. The shift in priorities evinced by Shea's advice indicates that engineers came to recognize more fully the rhetorical and enunciative specificity of different modes of sound production and the types of realism appropriate to each. Thus, an alternative ideal—one based upon rhetorical, narrative, and diegetic functions as well as upon narrational clarity—competed with the model of the invisible auditor.

While this new tendency might indicate that the "gut reaction"[35] of the average technician was changing, merely admitting that one might—at times—do something other than what was strictly allowed by the fidelity model did not mean that technicians had already developed a consistent logic guiding those deviations. Since soundmen did not want to believe they were applying different standards willy-nilly, they sought to validate the implied transformation of standards and practices by developing a new conceptual justification and logic of representation that agreed with both their new instincts and their new corporate responsibilities. This they found in another professional group, the cinematographers of the ASC. In a 1934 article entitled "Getting Good Sound Is an Art," Harold Lewis draws a parallel between the photographer who purposely over- or underexposes a shot, therefore departing from the "straightforward commercial ideal," and the sound recordist charged with recording film sound.

> Dramatic sound-recording must in the same way often depart from the standard of the commercially ideal recording. Like the cinematographer, the Recording Engineer must vary the key of his recording to suit the dramatic needs of the story and scene.

That this needed to be stated explicitly indicates that, as late as 1934, the "commercially ideal recording" (in other words, the perceptual fidelity or invisible auditor model) still exerted its force over the mindset and professional standards of sound recordists. In contrast to the musical recordist who always seeks to record the nuances of specific spaces and performances with the greatest perceptual fidelity, the film recordist "must know how each scene fits into the pattern of the picture as a whole, what precedes it and what follows, so that he can give it the best and most dramatically expressive aural treatment possible."[36] These particular thoughts carry an added weight because the writer was the president of the newly formed Society of Sound Engineers. The society's stated purpose was to bring the goals of sound engineers and the film industry into accord and enhance the engineers' prestige within Hollywood, much as the American Society of Cinematographers had done for its members.

The importance accorded to knowing "how each scene fits into the pattern of the picture as a whole" implies that engineers, revisiting the dilemma musicians had faced a few years earlier, had recognized that the classical film is better understood as a "pluripunctual"[37] unity—a combination of fragments whose unity is constituted primarily at the level of their combination rather than as a collection of individual, essentially discrete, and autonomous units. An aesthetic that preaches the absolute duplication of perceptually discrete, profilmic events will place far too much emphasis on the texture and idiosyncrasies of the individual shot or take by giving theoretical and practical primacy to the prerepresentational event rather than its diegetic representation.

The ideal of perceptual duplication presupposes, as I've argued, that the acoustic specificities of the original space and performance are necessarily and directly related to those of the finished representation, so that there is, conceptually at least, no distinction between the spaces of representation and reception—the ultimate movie audience is implicitly present on the set, and the finished film attempts to duplicate the experience of that observer. Yet, as we know, the classical cinema institutes a decisive break between these spaces, with the result that the represented space and time may have almost nothing in common with the actual spaces and times of production. The lack of intrinsic sonic hierarchy that the invisible auditor model implies assumes that the movie audience is, in essence, a collection of neutral observers with no preferences or expectations with regard to image and sound priority—they are simply witnesses to an untampered reality. This set of assumptions makes the connection of shots and takes into larger units extremely difficult, however, since the volume and quality of sounds may vary in any way between successive shots as a literalized perceiver flits about the set, following the eyes of a corporealized camera. Classical film, in contrast, thrives on the continual foregrounding of narratively important elements against a background of less important but generally "realistic" elements, and on a dissociation of camera and microphone narration from real perception.[38]

Without the sort of sonic hierarchy implied by the demand to keep dialogue intelligible, and to minimize the perceptibility of changing relations to each shot when the pieces are edited together, it would be difficult to construct a continuous, multiunit whole. Indeed, an excessively faithful recording can render continuity nearly impossible, since any scene with a continuous background sound will appear discontinuous if cut together out of different takes. Such difficulties encouraged technicians to record *only* the dialogue on the set, with a mike near the actors, adding characteristic background sounds later. This allowed continuity across cuts in the overall construction of films because discontinuous lines of dialogue (or other foreground elements) would nevertheless exhibit a consistent volume and intelligibility from take to take, thereby allowing them to be placed within an artificial, dubbed, "continuous" background. This technique granted the ensemble a feeling of wholeness and uniformity despite the actual discontinuities of production. In other words, the integrity, and therefore the idio-syncrasies, of the "pro-phonographic" performance were usually sacrificed in the service of a higher level of rhetorical or narrative continuity.

The case of background music makes this point in an obvious way. Under the model of fidelity, which stresses above all the integrity and unity of the original performance and the individual recording of that event, multishot musical sequences would have either to be filmed with multiple cameras and a single continuous sound take (as was the case for a short time)[39] or com-posed of various different performances of the same musical sequence. Since getting a single perfect performance was difficult and multiple-camera shoot-ing expensive and inefficient, such sequences required construction. However, piecing together a composite performance out of several "real" performances (as the invisible auditor model required) meant cutting both singer and orchestra at the same time for every edit. In a 1929 discussion of volume control, Carl Dreher points out that cinema audiences have not yet become sophisticated enough to complain about "synchronized pictures in which, as the scenes [shots] change, one musical selection is abruptly broken off and another starts will [sic] full volume in the middle of a bar."[40] Although I have been unable to locate a scene in which such an abrupt change occurs,[41] the example brings forward the importance of maintaining certain forms of continuity *across* cuts in the overall construction of films. It simultaneously sacrifices the idea that one needs to record rather than construct sonic per-formances. The musical performance is simply a highlighted version of a more general problem, though, since most scenes will appear discontinuous if edited together out of different takes if the narratively (and economically) mandated foreground is not accorded priority over the less important but entirely "natural" background.

Now, while such concerns and the adjustments of technique they would require might be acceptable to, say, a cinematographer, they flew in the face of the established norms and standards of sound engineers. Indeed, modify-ing theoretically prescribed microphone technique by, for example, moving it on a boom to follow an actor around the set (which soon became the norm in Hollywood) cannot readily be reconciled with either the practices of sym-phonic recording or with an "invisible observer" model of cinematic narra-tion. Falsifying the acoustics of the set or creating an imaginary acoustic space does not at all accord with the principle understanding of realism which shapes the engineer's self-designated technical brief—to record and thereby simulate as accurately and with as much nuance as possible, a sonic event from a stationary, audience-like position.

In point of fact, sound technicians did precisely falsify things again and again. As Maxfield pointed out on several occasions, it did not seem a professional imposition to alter the acoustics of the set in order to produce, through the taking of "a good commercial record," a recording that had the characteristics (in terms of reverberation, for example) desired by filmmakers. If a space "really" had acoustics that were deleterious to speech intelligibility, it was deemed acceptable to change the acoustics so that a "normal" recording remained intelligible. Perhaps because set construction was not obviously the responsibility of the recordist, it was easier to justify what might otherwise legitimately be called a "deviation" from the ideal.[42]

Of course, such foregrounding or prioritizing techniques were commonplace in concert hall architecture and thus already within the realm of the acceptable. That concert halls are acoustically unlike any other spaces, and are therefore reasonably considered acoustic devices, seems not to have entered the practical equation, as it did not enter the theoretical one, because such spaces are an integral aspect of music as a social and cultural practice. Had engineers made it theoretically explicit to themselves from the outset that acoustics were simply another instrument in the sound recording chain, and had not the bulk of acoustic research and practice been guided by the heuristic of the real, situated listener, perhaps it would not have seemed a deviation to shape film sound representations in other ways.

Given two recorded sounds with essentially the same acoustic characteristics, it would be impossible to determine whether an absence of background noises (behind, for example, a voice) was the result of elaborate soundproofing or of closely miking the actor with a movable boom. Thus, the finished products might be absolutely interchangeable. However, engineering standards of fidelity and representational correctness admitted one technique but not the other. Recording a sound event that was already hierarchically ordered according to a particular social function allowed the engineers to maintain a "hands-off" descriptive aesthetic which eschewed any overtly narrational or rhetorical use of the sound apparatus, while nevertheless achieving rhetorical effects. The "untampered" original was *itself* already a rhetorical device.

The emphasis on recording or inscription didn't completely solve the problem and was, in fact, something of a hindrance. Sound technicians repeatedly needed to reconcile themselves to a mode of production based not on the faithful collection of real events, but rather the construction of carefully hierarchized events whose realism is a function of their plausibility and their compatibility with conventions of narrative realism rather than literal, perceptual duplication. Part of the difficulty technicians had in recognizing these facts was the widespread tendency to delimit arbitrarily the processes of representation to the simple act of recording. Commentators from many different perspectives shared a tendency to treat effects of filmic representations as if they were causal factors or as if they were conditions that existed independent of the processes of representation. A. Lindsley Lane describes his understanding of the ultimate goals of cinematic representation, through the notion of the camera's "omniscience," and suggests that motion picture technique should as much as possible efface itself, since a picture which

> gives self-evidence of its making is not a good picture artistically and holds the chemistry of dissolution within its own structure, drawing the audience's attention away from its story-experience purpose; is, in other words, destructive to intactness of the "illusion of occurrence," which illusion is the psychological key to a successful motion picture-percipient experience.[43]

Lane here recognizes that the *goal* of cinematic representation is to create the "illusion of occurrence," that is, the represented *effect* of an event that seems to exist independent of the act of recording—an event that seems simply to "occur" and to be captured. This contradicts the writings of, say, Maxfield, who, seeking the same effects, still takes the idea of capturing (or "recording objectively") as a necessarily *actual* process and therefore (like those early cinema producers who manufactured "real" occurrences) manipulates the original through staging in order that it *can* be simply "captured." Lane sees the effect of the captured as precisely that—an effect. His insight seems not to have been shared by many of his contemporaries, although those who failed to share it should not be considered naive. In our day the belief that there is such an all-important, independent, prerepresentational event manifests itself, for instance, in the privilege accorded the original sound in academic theory. Such tendencies in both informal and formal theory tend to neglect the extent to which *all* aspects of scenography can and should be understood as participating in narration—not just filming and editing.

The role of the cinematographer as a model for the sound engineer therefore takes on even more importance since his prestige in Hollywood was unargued and his standards and practices perfectly suited to studio needs. By reimagining their role as similar to the cinematographer's, sound engineers ultimately jettisoned the all-defining "original sound" and the aesthetic of duplication it entailed and, for better or worse, set about the business of manufacturing sonic worlds whose parameters were judged mainly in terms of their internal coherence and representational functions.

Thus, by adopting the cinematographers' justification for manipulating images and the prestige of his "artistic" deviations, engineers shifted their emphasis from producing series of discrete and autonomous units toward producing takes with an ear toward their place within a larger series of representations whose combined unity and continuity took precedence over that of the individual representation. The norm of the invisible auditor gave way to the ideal auditor. Along with this shift came a redefinition of the standards of accuracy held to be constitutive of "professional" technique. By redefining the requirements of professional identity, sound technicians accommodated themselves to the reigning norms of the Hollywood film industry and thus retained not only their standing as good engineers but also paved the way for their acceptance and advancement within that industry. So, what might otherwise seem a rather marginal debate over representational theory can be understood as a crucial element of the transformation of a large, capitalist industry. Only after sound technicians conformed their standards and their senses of professional identity and success to standards compatible with the success of their new corporate employers, could sound engineering become an integral part of the Hollywood system. Representational theory was the primary ground upon which this realignment, redefinition, and integration occurred.

CHAPTER 6

SOUND SPACE AND CLASSICAL NARRATIVE

If, ultimately, the sound engineer accommodated himself and his standards to the classical Hollywood paradigm, and thus to capitalist rationalization, the process was not without its hesitations, discontinuities, and disruptions. Too often examinations of Hollywood's methods of representation assume as self-evident that the classical paradigm's norms were predetermined by the apparatus's built-in ideological predispositions. Crucial questions about the classical construction of space, for example, are answered as if by simply illustrating a similarity between filmic images and Renaissance paintings we prove their mutual implication. This approach, widespread as it is, not only mistakes essential elements of a Renaissance mode of representation, but through the language of legacies and inheritance[1] in fact *naturalizes* the emergence of modern representational norms by locating them in the structure of the technology itself.

Although Hollywood largely succeeded in its (often unconscious) attempts to control as many aspects of film production as possible, and managed to insinuate its needs and standards into the assumptions and practices of self-designated scientists, the accomplishment was no preordained certainty guaranteed by the grinding of lenses or the psychic structures of the bourgeois subject. The Hollywood mode of representation ensured its dominion much more subtly and pervasively by encouraging individuals and groups to accommodate themselves to and finally internalize its own historical norms in the guise of their own scientifically and aesthetically derived necessities or standards. Replicating a process already well known in industries around the country, Hollywood's engineers adopted the goals of capital as their own and came to equate corporate success with personal success. If we are to understand how this (by no means inevitable) process occurs, we need to examine the establishment of the most basic representational norms in their full complexity. The basics of Hollywood sonic space construction offer one such arena of conflict and accommodation.

The competing models of sound representation that shaped both filmmaking and standards of professional practice during the transition era are not simply theoretical or heuristic in nature. To reiterate what we have seen in chapters 4 and 5, each model implies and organizes a different form of filmmaking practice, and each suggests a distinct basic understanding of filmic space. The first, heir to metaphors of human simulation and described in terms of perceptual fidelity, emphasizes the literal duplication of a real and embodied (but invisible) auditor's experience of an acoustic event. Its watchwords are presence and immediacy, its standard, a fully textured and self-sufficient original sound completely autonomous of the act of recording. Aesthetic perfection entails the absolute re-presentation of the original in its fullness so that no differences whatsoever exist between the pro-phonographic experience and the phonographic one. Notionally at least, the auditor is literally made present at the (necessarily) prior acoustic event in such a way that all mediation is effaced.

The second model traces its roots to the metaphor of writing, and emphasizes the mediacy, constructedness, and derived character of representation. Its watchwords are repetition, legibility, and nonpresence, and its standard the impression it makes upon an auditor irrespective of the nature of the original. This model is concerned far less with duplication than it is with production, less with fidelity to the original than with citing bits of various originals to particular representational (and often new) effect. Rather than self-sufficient and autonomous, the original here is always determined by the nature of its mediated representational offspring. It has no particular significance in itself, but is rather the first (and unavoidable) step in a chain of events leading to a finished representation whose characteristics are governed by the demands of a particular representational goal and institution. Like writing in Plato's sense, these repetitions may take on or produce meanings profoundly at odds with their originals—indeed, adherents to this model count on this fact in order to create seamlessly coherent diegetic worlds out of the combined fragments of many "real" worlds.

Theoretically, and one might also say ideologically, each model implies a different relationship between its auditor and its represented world. The fidelity or simulation approach assumes that sound and space are homogeneous, without given, internal hierarchies of importance—phenomenally replete. Since it assumes that correct representation entails the complete duplication of every aspect of the sound in question, and that every aspect of that sound is equally essential to its identity, it privileges the original sound event over all other considerations. Consequently, it attempts to unite the spaces of reception and representation—to place the auditor as literally as possible *in* the profilmic space, and thereby literalize represented space and simultaneously efface representation in favor of unmediated presence. Of course, this goal is unrealizable and was recognized as such, yet it functioned, all the same, as a regulative ideal. This approach similarly assumed a very specific model of perception—one that may, itself, have found its model in the microphone. Perception was understood as ideally sensitive and "passive" in the sense that it imposed no order upon sounds and assumed no listening goals other than sensitivity.

The alternative model, in contrast, presumes that both sound and space are heterogeneous, structured differentially by hierarchies of importance. Since perception is understood as directed and goal-oriented, sounds are ordered by their relevance to the desires of the listener. Hence, there are important and unimportant sounds, and sonic tiers or "layers." Represented space can therefore be fabricated, with emphasis directed toward maintaining and foregrounding desired sonic factors at the expense of less desirable ones. Sound space also need not be real—it is constructed to produce a particular effect (often "realism"), and to be plausibly, rather than literally, coherent. Unlike the fidelity model, which necessarily assumes that volume, frequency range, timbral quality, scale, and so forth are absolute values, here all sonic attributes are measured purely in terms of internal consistency. Rather than unchanging and autonomous concepts, defined by neutrally descriptive terms, perception, sound, and space are treated as closely interrelated—even interdependent—phenomena, contextually determined by the demands of a particular representational system.

We can see the difference between these two models if we look at the different meanings assigned to the word *perspective* in the transition era's debates about film sound space. The fidelity approach assumed that the spatial perspective of the sound event should match that of the image. As Joseph P. Maxfield, one of the most prominent sound technicians of the early sound era and head of Bell's Electronic Research Products Incorporated (ERPI) division, which was responsible for developing and installing Vitaphone equipment, argues, "The technic of acoustic control is based on letting the camera be the eye and the microphone the ear of an imaginary person viewing the scene."[2]

On the other hand, perspective might also be modeled on the Hollywood image. Such a space, David Bordwell argues, was comprised of a "foreground" and a "background," ordered by their contributions to the narration.[3] Classical sound "perspective," or mise-en-scène, created discrete and carefully separated "planes" of sound in imitation of classical image space. While this may be true in the long run, it ignores and elides a significant period of controversy surrounding the nature of sound space. Consider this 1926 statement from Maxfield:

It is customary in the phonographic art . . . to increase a solo part to a level of loudness somewhat greater in comparison with its musical accompaniment than would be the case in an actual stage performance. Similarly, in the moving picture field, it has been the custom to hold the attention of the audience on the important features of the picture by keeping them in the foreground and considerably larger than life size. The first audible motion picture was made in accordance with these two standard conventions, and the result was most disconcerting. The eye and ear of the observer obtained, as to the relative location of the various artists, separate impressions which differed sufficiently to make it impossible for the observer to focus his attention on the performance.[4]

In other words, Maxfield and his colleagues initially assumed that sound events, like the visible world of the classical cinema, possesed an intrinsic internal hierarchy (the solo more important than the accompaniment, the actor more than the decor), and that representational scale should be determined by the demands of audience interest. These "standard conventions," however, appeared incoherent in the Vitaphone film, Maxfield claims, because sound scale and image scale appeared not to match. Typical of the technicians of the era, he *rejects* the norms Bordwell finds in later films in favor of the idea that films simulate the perceptions of an observer located on the film set, whose eyes and ears (camera and microphone) are joined as inseparably as those of a real head.

The later pictures, in which the sound was picked up as it would be by the ears of a listener to the original performance and in which the picture left the images in their natural sizes, removed these difficulties and produced

an illusion so satisfactory that many of those observing it lost themselves
in the performance and forgot that they were listening to an audible picture.[5]

Moreover, it is symptomatic and utterly consistent that Maxfield's implicit
standard assumes an absolute sense of spatial scale—the human body as we
experience it in the "real world." It therefore also assumes that representation
is merely a matter of duplicating the experience of a profilmic witness as if
the film audience and actor were copresent, sharing the same physically continuous space.

However coherent as a theoretical approach to sound, the surrogate perceiver model would soon prove inadequate to the classical cinema's narrational demands. The paradigm was not, however, immediately replaced by
the more analytic and narrative approach identified by Bordwell. Instead,
"fidelity" continued to shape how technicians approached basic practical
problems regarding synchronization, sound location, and editing, among
others. When Maxfield, for instance, is bothered by a contradiction between
the apparent locations of a sound and its visible source, he does not simply
suggest we link the surrogate eyes and ears to an imagined body, but he insists
that the *represented world* be presented *life-size*. Rather than adopt another
strategy, Maxfield applies his perceptual standard even more literally. And
what was true for Maxfield was true for the rest of the field.

In his own analysis of sound space, Rick Altman persuasively argues that
despite all this worrying over absolute perceptual fidelity and careful matching of vision and sound, by 1938 Maxfield himself could, in a complete reversal, argue for uniform close miking regardless of camera distance, indicating
that the technician's "gut reaction" had shifted decisively in the intervening
years.[6] Altman goes further, arguing that "it is no exaggeration to claim that
this reversal represented a fundamental turnabout in human perception."[7]
Even if perception "itself" could not change in the space of eight or ten years,
the cultural value of a particular form of perception and its role in a broader
representational system certainly could. So, while *hearing* may not have
changed, the dominant and regulative mode of *listening* had. Pressures both
representational and economic encouraged technicians to adopt a new set of
norms, but the inconsistent application of different strategies and standards
indicate that despite Hollywood's needs, the sound engineer's ingrained, professional disposition toward representational issues delayed their arrival.
Transforming "gut reactions," it turned out, involved radically redefining the
dominant understanding of sound space and consequently the nature of

sound itself. Presaging this change, a variety of related cracks appeared in the once uniform facade of perceptual fidelity.

Underscoring Maxfield's concerns about linking sounds with their pictured sources, other writers turn less to real perception than to audience effect. For example, in 1928 H. B. Marvin notes that

> the chief requirements of a good system combining sound and pictures are: first, synchronism between sound and picture; and second, distortionless recording and reproducing; that is a high degree of faithfulness of sound both in loudness and quality.[8]

While still insisting upon fidelity, he concedes that synchrony is just as important, citing the example of "stage sound effects where the primary object is accurate timing and dramatic effects, and the primary requirements are synchronism and adequate volume of sound."[9] By pointing us to the world of coconut shell horses and tin sheet thunder, Marvin indicates both that one may artificially link a sound and a source, and that sounds may fulfill different functions. In Marvin's analysis, "dramatic effects," in fact, seem to require *only* synchrony and volume. Spatial literalism and timbral fidelity are simply not an issue. Other writers offered similar opinions over the coming months, and even Maxfield would come to modify his basic definitions of sound, space, perception, and representation.

In another foundational essay, Maxfield offers that the goal of the sound motion picture is the "pleasing illusion of reality."[10] Said illusion is attained, he claims, by duplicating as nearly as possible the conditions of real, fully sensitive binocular and binaural perception. Beyond his predictable advocacy of absolute fidelity, however, he proposes two important and transformative qualifications. First, he counsels that a recording need only control the ratio of direct to reflected sound in order to produce a convincing illusion of depth, and second, he cautions that "too much reverberation can overemphasize incidental noises."[11]

By acknowledging that, for monaural recording, the ratio of direct to reflected sound is more important in determining spatial location (or at least distance from the camera) than actual miking distance, Maxfield effectively relinquishes the concept of strict perceptual fidelity. Likewise, the idea of "too much" reverberation cannot be reconciled with the fidelity position. If one records a sound faithfully, the reverberant properties of the room are simply given, inseparable from the sound "itself," and unalterable. The fidelity model, which defines both audition and recording identically, makes no

such distinction between intrinsically important and unimportant or incidental sounds—all aspects are uniformly significant. And yet, Maxfield criticizes recordings with "too much reverberation" because they overemphasize
"incidental noises."[12] In a way he appears not to recognize, Maxfield's com-
monsense hierarchy touches the very core of sound itself, radically but qui-
etly transforming a unified and homogeneous event into a heterogeneous
and stratified structure whose parts may serve different and even incompatible functions.

While Maxfield's distinction between incidental and important sounds
should therefore have invalidated the assumptions of the fidelity model,
technicians continued to devise techniques that emphasized *some* sounds at
the expense of others, but still appeared to respect the ideal of fidelity. Rather
than increase the volume of the desired sound in relation to the "incidental
noises" or move the microphone closer to the main source than the camera
position implies (which obviously would have violated fidelity principles),
Maxfield eliminates incidental noises by excluding them, and reduces reverberation by damping the three-walled, ceilingless set. T. E. Shea, also an
advocate of soundproofing, even goes so far as to claim that "the stage is the
first, and the only acoustic element in the [recording] channel,"[13] thereby
contradicting the tenet that one simply records a preexisting and wholly
prior original sound by including the "original" in the very process of manipulation. This elision, however, tallies perfectly with the tendency among academic theorists and others to regard the architecture of concert halls as
"neutral."

Since, as Maxfield argues, the perceived location of a sound source
depends primarily on the ratio of direct to reflected sound, manipulating set
acoustics effectively provides a functional equivalent for *actual* microphone
distance from the source. That is, by manipulating reverberation, technicians
learned, sounds might be "placed" at any distance from the camera. Thus,
Maxfield and Shea, like sound engineer K. F. Morgan before them, had discovered but not fully admitted that the filmic effect of sound sources was just
that—a *constructed effect* with no simple and direct relationship to a real state
of the profilmic world.[14] However, the fact that the construction materials
and design of the set could influence the character of the original sound paradoxically offered Maxfield and others a way of reconciling the need to
reduce reverberation while maintaining the fiction that they respected the
priority of the profilmic arena.

The distinction between desired and incidental sounds found its place
within a larger set of distinctions and hierarchies required by the demands of

Hollywood narration. From 1932 on (and probably as early as 1930) film sounds were conventionally divided into hierarchically ordered tracks, with "dialogue" assuming the leading role. This classification, however, is neither purely objective nor neutrally descriptive—it is determined by and attuned to Hollywood categories and imperatives. Dialogue, for instance, is not the unambiguous category it may seem, since comments like "whenever *effects or dialog* are present on the dialog track"[15] made it clear that "dialog" designates a *function*, not an object. In this case, "dialog" does not signify only speech, but rather the sound assumed to be most important to the narration—that is, the sound "speaking" the narrative at a particular moment.

In Howard Hawks's *The Dawn Patrol* (1930), for example, an Army Air Corps ground crew await the return of their aviator comrades, silently counting the number of planes they hear passing overhead in order to determine how many have returned from their perilous missions. In this scene the sound of planes may be said to occupy the "dialog" track because it asserts its functional primacy. Similarly, in John Stahl's *Imitation of Life* (1934), the opening sequence is dominated by a succession of offscreen sound effects that illustrate how Claudette Colbert's household, and indeed her life, have gotten beyond her control. Her child, a telephone, a boiling coffee pot, a shout, and a knock at the door all compete for her attention and draw her through the house as she tries to manage both her home and her business. In both cases, effects are more narratively important that the incidental words spoken at the same time, and are therefore foregrounded in volume and intelligibility.

Edwin Krows's 1931 essay, "Sound and Speech in the Silent Film," locates the rationale for thus redefining sound in the difference between sound as phenomenon or material and sound as signification or function. With this idea in mind, Krows advises us "never to think of the spoken word *merely* as sound. The voice makes a sound, but the word uttered may express the reverse of the sound."[16] When suggesting relevant precedents for representing the film voice, this functionalist attitude turns its attention not to gesture or music but to title cards, thereby aligning speech with intelligibility and writing over fidelity and perception. In a similar move, John L. Cass compares sound recording with a cheap but readable book, stressing how even here "incidental" material features may impede comprehension.

> The illusion thus created [by the book] depended on the fact that legibility was the one requirement of that particular medium. . . . On the other hand, graceful type which was more difficult to decipher would have detracted from our pleasure, as it would have made the act of reading more

difficult, thus thrusting the medium upon our consciousness, when that consciousness desired to be alone with the meaning of the printed words.[17]

Presuming that "our consciousness desires to be alone with the meaning" of sounds and images as well, he argues that, "As in the case of the printed word, the pictures must be easily discernible, and the sounds easily understood."[18] As Krows claimed, it is a mistake to think of speech as mere "sound," since in a film a recorded sound is not simply an acoustic duplication but an element of a larger signifying system. Like a florid typeface or handwriting, a recording that maintained a literal allegiance to the rhetoric of "fidelity" would record every sound indiscriminately, oblivious to confusing echoes, narratively incidental sounds, or camera noise. Consequently, Maxfield, Krows, Cass, and a growing army of technicians believed that sound could no longer be defined "objectively" by purely physical criteria, nor was it internally homogeneous, but like speech, it possessed some attributes that, within the domain of a particular cultural *function* (however that was defined), were more important than others.

The emerging hierarchy between essential sounds and mere noise grew in authority throughout the 1930s. Concurrently, it became the norm *not* to match visual and acoustic "scale," *not* to locate the microphone with the camera, *not* to respect the acoustics of the space of production, and *not* to offer a perceptually based "coherent point of audition" with which the spectator could identify. Instead, technicians developed a flexible set of norms that sought to enhance intelligibility through close miking and sound-proofing.

This distinction between different types of sound altered the course of technological development, too, and a steady stream of sound innovations bore its imprint. Harry Olson and Irving Wolff announced the development of a parabolic horn "sound concentrator" for microphones in April 1930, Carl Dreher an improved version of a similar device in January 1931, Olson a new mass-controlled electrodynamic (ribbon) mike with "marked directional characteristics" six months later, and Weinberger, Olson, and Massa a "uni-directional" mike useful for increasing the direct-to-reflected sound ratio and eliminating camera noise in 1933.[19]

Quieter and increasingly more mobile microphone booms offered the possibility of following characters around the set, enabling recordists to capture a high degree of direct sound and, therefore, dialogue intelligibility. The boom allowed actors to retain the kind of mobility they had been accustomed to in the silent era, while keeping the microphone within a few feet of the speaker, and rendered the quality of the recorded sound effectively "independent of the reverberant properties of the set," as one early report put it.[20] (See figure 6.1.) Responding to the same pressures in another way, the *Journal of the Acoustical Society of America* for the period is a gold mine of information on how, contrary to the trend in concert hall acoustical research, one might best *reduce* the reverberations created by materials present on the movie set.[21] As Rick Altman argues, each of these research trends had a similar end in view: to maintain dialogue intelligibility.[22] What is more, each development gave concrete embodiment to a representational norm—by instrumentalizing it, making it automatic, rendering it part of the "basic apparatus."

As it turned out, however, the answer to Hollywood's demands could not be to seek intelligibility at all costs. The filmmaking techniques implied by these developments highlighted another dilemma, identified by Franklin L. Hunt in symptomatic fashion in the use of mobile booms:

> To retain the advantage in articulation of the close-up microphone and at the same time permit the freedom of dramatic action required, the microphone is sometimes moved about on the stage, that is, "panned," to keep it always near the speaker as he moves about while the record is made. This is done at the risk of naturalness and of the completeness of the illusion that the voice comes from the speaker as he appears on the screen.[23]

Since Hollywood sells not only narrative pleasures but a form of realistic representation, a form of technical and formal perfection, and an array of spectacular effects, Hunt seems to argue, absolute intelligibility is not always the appropriate criterion by which to make practical decisions. Just as there were different kinds of sounds (dialogue, music, sound effects), there were different functions as well, some of which did not always complement intelligibility.

In fact, intelligibility always functioned in tandem with the reciprocal ideal of fidelity. RKO's Carl Dreher suggests, in fact, that one cannot do without the other: "Broadly considered, naturalness of recording and reproduction might be considered the sole aim of sound technic. From this viewpoint, naturalness would require intelligibility at least as good as in the original spoken rendition."[24] "Intelligible to *whom?*" is the obvious question here, since for someone far away from the original source, unintelligibility might be completely "natural." Dreher nevertheless raises an important point. It is never *strictly* intelligibility that is at issue, since sounds, even language spoken over

FIGURE 6.1

On the set of *Girl of the Golden West* (1930). Using a bamboo microphone boom to maintain "close-up" dialogue characteristics. (© 1930 Turner Entertainment Co. A Time Warner Company. All Rights Reserved)

a telephone, signify in ways other than purely linguistic ones. As Bell engineers W. H. Martin and C. H. G. Gray point out regarding telephone transmission standards:

> From the standpoint of telephony, the major importance of a difference between the original and reproduced sound is determined by its effect on intelligibility. . . . The tolerable departure of the reproduced from the original sounds is limited also by certain effects which are noticeable to the listener *before* they materially affect intelligibility, such as loss of naturalness.[25]

From the perspective of science just as much as Hollywood, the pair *intelligibility* and *naturalness* are intimately related, although (I have argued) mediated by different understandings of perception. Even the telephone, which is defined almost completely by its communicative function, requires at least a measure of naturalness.

However imprecise the term *naturalness* seemed, providing a practical working definition proved not to be a difficult task. Hunt even identifies one of the specific sonic features constitutive of "naturalness":

> It has been shown that the per cent articulation [intelligibility] in a room decreases as the distance from the microphone to the speaker increases. . . . The best articulation in sound picture recording is obtained with the microphone within a few feet of the speaker but . . . [this] leaves something lacking when used with middle distance or long shots. The *unnaturalness is due to the absence of the sound reflected from the walls* of the set which *when added to the direct sound* from the speaker's voice *simulates the quality obtained in ordinary rooms under listening conditions.*[26]

Hunt, in effect, reconceptualizes the opposed terms by implying that reflected sound is something extra that can be "added" to the direct sound as a purely contingent feature may be added to a substance without altering its essence. Although physically inseparable, direct and reflected sound are analytically distinct and can perform different but complementary functions. In short, reflected sound may function as something like a diacritical mark—an added signifier that alters a sound rather than an essential component of the sound "itself." By separating a now inherently heterogeneous sound into different functions, we get a glimpse of how technicians were to redefine perception, too.

Despite Hollywood's widespread use of mobile booms,[27] one commentator suggests that, theoretically at least, they still pose representational dilemmas. Franklin Hunt warns that technicians use booms for close miking "at the risk of naturalness and completeness of illusion."[28] A few months earlier however, the need for close miking was "quite well known." The newly recognized threat entailed a loss of "naturalness," which by this point had come to be equated with reverberation—a *background* property.[29] In contrast to the unified fidelity approach, the discontinuous, *additive* foreground/background model of space is in no way threatened by close miking, just as the "blue-screen" approach to image space is not threatened by focusing on foreground characters against a blue background. The background is not *lost*; it is simply not yet *added*.

John L. Cass, whose analogies between sound recording and printing argued that intelligibility was the key concern of sound representation, still felt compelled to turn back to perception as the final arbiter of spatial construction, regardless of other standards. Cass argues against using more than one mike for any scene, even to enhance legibility, because it violates the integrity of the imagined body:

> When this scene is projected, the eye will jump from a distant position to an intermediate position, and from there to close-up positions on important business. The sound will run throughout as though heard from the indefinite position described above. Since it is customary among humans to attempt to maintain constant the distance between the eye and the ear, these organs should move together from one point to another to maintain our much mentioned illusion.[30]

If, finally, sound recording was still destined to be measured against the standard of perception, the reigning definition of perception itself had to change. However, given the demands of Hollywood narration, this would be somewhat difficult. In negotiating the relationship between fidelity and intelligibility, technicians resort to locutions which today sound strange, to say the least. In a 1931 discussion of dubbing, for instance, George Lewin argues against shooting and recording on location. After noting that the microphone, while nearly as sensitive, is far less selective than the human ear, Lewin claims that "a microphone, unfortunately, is not capable of differentiating between *what we are trying to record and the background noises*, for it will pick up the latter *with discouraging fidelity*."[31] Oddly, "fidelity" (and an explicitly perceptual version of fidelity at that), which had previously been the sine qua non of mechanical perception, is here an object of consternation. In short, real perception had become problematic.

Harry Olson and Frank Massa, in contrast, offer a simple rhetorical adjustment that brings perception into alignment with the selective sensitivity of
directional microphones. After setting out the familiar litany of properties
(complete frequency range, complete volume range, complete reverberant
and binaural characteristics) assumed to be necessary for realistic sound
reproduction, they suggest that microphones with a directional sensitivity
that *cannot* preserve frequency, volume, or reverberation in their "natural"
proportions offer a kind of functional equivalent to the act of mental attention or auditory discrimination. This feature, they claim, allows talented technicians to create "a sonic 'center of gravity' comparable to the 'center of
gravity' of the action."[32] Wesley C. Miller had implied as much several years
earlier, noting that a natural effect is the result of hard work.

> We create artificially the effects of stereoscopic photography, we exagger-
> ate perspective or we attract attention to some person through the device
> of the close-up. Each time as we do this we keep in mind the probable
> effect which is to be produced in the theatre. . . . With sound our problem
> is identical.[33]

"Exaggerating" in order to "attract attention." Artificiality in order to produce the *effect* of naturalism. Directionality to produce the *effect* of attention.
The emergent theoretical model implied here differs in important and, as far
as Hollywood film production was concerned, crucial ways from the standard of absolute fidelity, insofar as it recognizes representational effect as
more important than strict perceptual literalism.

Rather than treating, say, Dreher's five-eared spectator as the desperate last
gasp of an ideologically overdetermined attempt to preserve the illusion of
subjective homogeneity in the face of its actual fissure, it is helpful to consider the possible usefulness of such an absurd image. On the one hand, its purpose is painfully evident. Film is a medium that presents complex perceptual
experiences for its audiences by aligning the spectator's eyes with the camera
operator's, and his/her ears with the recordist's. Thus it is hardly strange to
conceive of filmic representation as attempting to provide a simulated perceptual experience of a real event, and therefore to use perception as a standard. Any technique deemed unsuccessful might reasonably be explained as
deviating from normal perception.

From the technicians' point of view, these various perceptual heuristics
might also serve as professional criteria of value and appropriateness. The
straightforward invisible perceiver model allows unsuccessful practices to be

explained away without relinquishing the right to determine what counts as "good" practice. The malleability of the perception-based standard advanced by sound engineers offers apparently consistent grounds for disallowing particular (unsuccessful) techniques while accepting others (successful) that would otherwise seem to depart from the model's expected limits. By introducing modifiers such as "mental attention," "binaural discrimination," "expectations," and "appropriateness," devices like directional microphones could be described as conforming to a perceptual model. Perception, the ostensible standard, was thereby redefined in strategic ways, suggesting that perception can never really serve as representational standard unless a particular *mode* of perception is specified.

Rather than impractically adhering to standards appropriate to symphony recording, the new standard focused less on exactly duplicating the pro-phonographic and concentrated on building a diegesis and producing spectator effects. Instead of perception pure and simple (what E. H. Gombrich has debunked as the "innocent eye"), the guiding metaphor became something akin to what A. Lindsley Lane called the camera's "omniscient eye"; it is a short leap from the notion of visual omniscience to its sonic equivalent. In Lane's terminology, the desired effect was to create the sense "of being at the most vital part of the experience—at the most advantageous point of perception" throughout the picture.[34] Here professional and artistic correctness are explicitly associated with a particular narrative effect—omniscience—rather than an abstract standard of absolute fidelity.

Given the various pressures to abandon the perceptually faithful model of sound recording, it remains puzzling why it had such staying power—rearing its head again and again in articles that seem to eviscerate its central canons. Were it *simply* a theory, it would have withered as quickly, but it was never just, or never simply, that. The fidelity model found support and elaboration not only in theory but in technicians' habits of thought and practice, in the research priorities of the largest laboratories, and in the very form of many sound devices, which were materially resistant to the demands of emerging representational norms. Most importantly, it found validation in an important early sound film form—the Vitaphone short—whose representational needs meshed seamlessly with the perceptual model of recording.

As is now well known, the earliest successfully mass-produced sound films, Warner Brothers' series of Vitaphone shorts and features, were not expected to be synchronized dialogue dramas. Instead, Warner hoped to profit by providing even the smallest local theaters with grand orchestral acompaniment for its features, and with the best in New York vaudeville acts as prologues.[35] The Vitaphone disc-based system of recording and reproduction was developed with this goal in mind, and in many ways reflects its assumptions, priorities, benefits, and limitations. I emphasize the shorts over the features here because they relate more directly to the techniques of classical sound cinema of synchronized speech, music, and effects than do the synchronized "silent" films that were their contemporaries. Despite their tributary relationship to the features, as "workshops" where the basics of sound and image recording were developed, their formal properties—especially their use of sound—indicate that they are not simply models for later films. The shorts, in fact, belong to a mode of representation significantly different from the norms of classical cinema—one dedicated to the absolutely faithful duplication of real acoustic perceptions.[36]

The classical mode of spatial construction emphasizes diegetic, multishot, narrative constructions whose primary compositional values are narrative saliency and intelligibility and whose spatial and temporal coherence are plausible, if imaginary. Space and time cohere only in a geography inferred by the spectator who is their point of coherence and their ultimate "cause." In this system, the spectator is conceptually, structurally, and practically primary to the practices of representation, with every decision therefore made with a view toward its effect on him or her. In effect, this system imagines the kind of "transcendental" spectator described and critiqued by apparatus theory.

On the contrary, the alternative, fidelity-based mode, including the Vitaphone shorts, is based on the spatiotemporal integrity of the prerepresentational sonic performance and assumes a situated and embodied—perhaps even empirical—spectator who is *part of* the profilmic/pro-phonographic space, although conceptually and structurally subservient to the performance. Representation, on this account, imposes no specific order upon the event, no hierarchies of importance, and assumes no specific spectator or auditor priorities. Actual profilmic relations to the position of an embodied and static spectator/auditor determine its spatiotemporal coordinates. In short, it assumes a different subject/scene relationship than does the classical mode[37] and therefore structures the probability of aesthetic decisions and practices in a different hierarchy.[38]

The Vitaphone system was based on the phonograph, taking advantage of Bell's existing research and patents in that area, so the "disc aesthetic" of perceptual literalism seemed perfectly appropriate. Musical records typically make no distinction between *represented* space and time and the space and time of the prerepresentational performance, since they routinely assume the fiction of an auditor who is actually "there" at the scene of performance in a definable and literally embodied location offering a perceptually specific (if conventionally chosen) "vantage point." It makes sense to think of this aesthetic as dedicated not to represented *sounds*, but to represented *hearing*. This general tendency was doubtless reinforced, as well, by the perception and economic positioning of the shorts not as movies with sound, but as illustrated records.[39]

Like the phonograph and radio industries from which it emerged, the Vitaphone short was institutionally and economically bound to an aesthetic that privileged an absolutely faithful rendering of a sonic performance conceived of as wholly prior to the intervention of the representational device.[40] As a practice, representation was therefore subservient to the spatiotemporal integrity of the original event, which was staged in its "normal" manner. A musical performance would be recorded in a concert hall and the microphone would attempt to duplicate the listening of an ideally sensitive concertgoer seated in the audience, thus preserving all nuances of performance, timbre, acoustics, and location. The spatial and temporal relations of the finished representation were assumed to duplicate those of the original performance, and the coherence of the composition was grounded in a unique act of perception, simulated by placing the microphone in a real audience.

Indeed, the static auditor position offered by Vitaphone's recording of Marian Talley's "Caro Nome" (from Verdi's *Rigoletto*), which causes her voice to diminish in volume and increase in reverberation as she exits the set, offers an instance where precise localization of the auditor in relation to the source enhances the overall effect of the scene.[41] It creates the vivid and dramatically satisfying impression of her retreat, just as it would have for an audience member in a concert hall. As is the case in most of the musical shorts, the performance is miked from a stationary and ideal position, interfering with the staging and performance of the sonic event only through the act of "listening," like an actual auditor. Moreover, the presentational style, frontality of performance, and mode of address typical of the shorts imply the presence of an audience *within* the space of representation which is here the *diegesis* as well.[42] Had she been speaking important lines of dialogue, such a decrease in volume and increase in reverberation might easily have been obtrusive, but here no special attempt is made to foreground the voice over the orchestra or even over the reverberation, since the reigning aesthetic demands that the sound performance "as it is" be duplicated rather than transformed for particular rhetorical or narrative purposes.

Spatial and temporal literalism, although effective here, could create enormous problems for classical scene construction. Changes in acoustics or in background noise from shot to shot, lack of guaranteed intelligibility, the

obtrusiveness of narratively inessential or unwanted sounds are all acoustically "real" elements that might impede the more general goals of classical
film—namely, the construction of continuous space and time and the efficient presentation of narrative. Why did these problems not present themselves in the shorts, and why did the logic of perceptual duplication still seem
creditable? In large part, because of the nature of the Vitaphone sonic performances themselves.

Vitaphone performances, especially musical ones, presented recordists
with "original" sounds that were, themselves, already rhetorical—that is,
structured hierarchically. When norms of composition, performance, and
architecture ensure that the desired sounds are foregrounded and those of
less conventional importance minimized, a perceptually literal aesthetic will
still produce an appropriately ordered recording. However, since conversations are not orchestrated, nor conducted in specially designed acoustic
spaces, a perceptual realism that subordinates representational interests to
the actualities of the event easily results in unintelligible dialogue and therefore a representation whose formal hierarchies are not suited to audience
expectations. These problems culminated in the eclipse of the Vitaphone
shorts as an alternative mode.

While the acoustic fidelity and subtlety of the disc were rivaled by no
other film-sound system, one could hardly have picked a system whose basic
technology was more poorly suited to the needs of the multishot, out-of-
sequence, spatially nonliteral, and industrially rationalized mode of diegetic
representation characteristic of Hollywood. There is even strong evidence to
suggest that Vitaphone engineers never even imagined editing a scene out of
separate takes, so different was their basic idea of representation.[43]

In his 1929 article "Scoring, Synchronizing, and Re-recording Sound
Pictures," K. F. Morgan offers an industrial and institutional explanation for
the delayed senescence of the fidelity mode—one that hinges on differing
conceptions of the relation between shot and scene. In a brief historical survey, Morgan points out that for early Vitaphone and Movietone pictures

the editing was fairly simple, *each take being one complete scene in itself.* . . .
The introduction of synchronized sound and dialogue into pictures of feature length presented the problem of sound cutting. . . . With the original
recording on disc, the cutting became a rather involved mechanical as well
as electrical process since the scenes as recorded had no definite chronological relationship to the final product. This introduced the first necessity
for re-recording sound.[44]

Morgan describes disc editing as a particularly laborious process. Operators stationed at each turntable *counted revolutions of each disc* in order properly to cue a series of separately recorded takes for combination in a particular scene.[45] The almost unbelievable complexity of this solution—which demanded the simultaneous labor of many trained technicians, and many hours to accomplish the simplest of tasks—as well as the technique's potential inaccuracy (how could one find an exact editing point on a *disc?*) indicate the remarkable extent to which editing, or the serial relationship of representations, was indeed secondary to faithful, real-time recording.

In a similar complaint, Nathan Levinson notes that until Al Jolson's *The Singing Fool* (1928), Warner Brothers' disc-based system physically precluded combining parts of different sound takes in order to create composite (multitake) wholes. Indeed, the very form of the device made editing almost inconceivable (see figure 6.2). Unlike sound recorded on film, discs could not be cut up into easily manipulated fragments that could then be arranged, rearranged, and combined into new wholes. For the Jolson film, however, Vitaphone technicians electrically connected a series of turntables that stopped and started in succession.[46] The image of a row of record players (with a separate technician attending to each) awkwardly turning on and off strangely embodies the conflict between different notions of the individual take, and its relationship to a larger series (see figure 6.3).

The earlier historical transition from single-shot films through films constructed as a connected series of relatively autonomous but chronologically ordered scenes, to a cinema of multishot scenes and narrative integration prefigures the development of cinema sound, which, in a similar fashion, moved both conceptually and technologically from unipunctual to pluripunctual forms. In *The Singing Fool*, the constructed *series* achieves a new importance, but the fragments of the series are still understood as discrete and autonomous wholes whose integrity is defined by an uninterrupted act of recording rather than *across* fragments, and whose very technology mitigates against their connection.[47]

Importantly, even editing strategies based upon multiple-camera shooting preserved this model because, while they allowed an editor to approach standard patterns of visual construction, they allowed no similar form of audio scene construction.[48] The continuity of the scene is predicated on the (actual, physical) unity of the sound take. While this practice often comes close to the feel of classical scene construction (present-day multiple-camera TV sitcoms seem fairly classical), it wasted film stock, personnel, and time, and imposed quite a number of arbitrary restrictions on staging and editing (mul-

FIG. Side View of Reproducing Equipment attached to "Simplex" Projector.
Upper Gear Box
Connecting Shaft
Lower Gear Box
Motor
Reproducer Mounting
Reproducer
Control Box
Record
Turntable
Turntable Pedestal

FIGURE 6.2
The Vitaphone projection system showing Vitaphone disc. (© Warner Bros., a division of Time Warner Entertainment Company, L. P. All Rights Reserved)

FIGURE 6.3

Because of the material form and limitations of the audio disc, mixing (combining) separate takes into a single sound scene was cumbersome and often inaccurate. (© Warner Bros., a division of Time Warner Entertainment Company. L. P. All Rights Reserved)

tiple-camera sitcoms rarely employ extreme close-ups, rapid editing, point-of-view shots, etc.).

Inadvertently and unhappily, under the multiple-camera/single-disc regime, visual editing was forced to become subservient to the uninterrupted unfolding of the continuous sound track, whose relentlessly linear and progressive development precluded interruption and recombination. In practical terms this meant that neither dialogue nor sound effects could be added, removed, or rearranged. The very tempo of a scene or even the timing of dialogue was inevitably defined by the temporality of the profilmic performance.[49] In short, the time *and* the space of the sound track were still *recorded* rather than *constructed*.

Morgan's account of disc editing indicates how ingrained but impractical assumptions about representation shaped even the initial research and production agendas of Bell and RCA. He points out that it was specifically

Hollywood's (rather obvious) need to *edit* that finally served as the motor for developments in the various rerecording and dubbing technologies, and that such developments would not have occurred otherwise.[50] That a sound industry so deeply invested in movie production should have needed the "wake-up call" provided by Hollywood's editing demands is surprising given the fairly obvious necessity of rerecording to any film understood to be comprised of multishot, or pluripunctual, scenes, yet this is just what the perceptual aesthetic misunderstood. The major labs were simply committed in economic, practical, and especially conceptual terms to the single continuous event–single continuous act of recording model of representation, with its consequent equation of shot and scene. Given the fact that every Hollywood film relied on editing, how could they have neglected its importance?

Why would Morgan find it "natural" that "combining sounds" was perceived as a "secondary adjunct" to recording them? His belief that "motion picture engineers will understand perhaps better than electrical engineers the necessity for . . . 'dubbing' "[51] indicates that the two groups' different emphasis on recording versus combining and constructing sounds derives from their different conceptions of the process and goals of sound representation in general. Morgan goes so far as to suggest that the inappropriateness of the disc-based system to the exigencies of editing hastened its demise as much as anything.[52] Geared toward the literal reproduction of musical performances, the disc system profoundly favored the *duplication* of single, integral events over the *construction* of events out of fragments. To oversimplify a bit, the disc model is defined by the microphone and the profilmic/pro-phonographic rather than by the editor and the diegetic.

As in the early cinema period, the most persistent spatial problem for transition-era image and sound was how to structure the individual recorded fragment so that it could function seamlessly as an element within a larger edited series or scene. In the case of the image, the problem was resolved through many of the techniques we have come to think of as definitive of continuity editing: narratively determined spatial hierarchies, selective focus within the individual shot, overlapping and contiguous spaces, eyeline matching, and match-on-action cuts *across the edits* between shots. Each of these techniques contributes in its own way to the overall effect of emphasizing the continuity of the scene or series over that of the individual shot.[53]

Segregating the image into a hierarchically ordered mise-en-scène allowed editors to construct a diegetically whole and apparently continuous space out of fragments of disparate origin. Foregrounded elements—usually characters—carried the weight of narration and determined editing. The softly focused, indistinct, generic, or unobtrusive backgrounds so characteristic of classical cinema help to produce a sense of spatial and temporal continuity and verisimilitude. In a functional sense, the background's main purpose is to situate the characters, objects, or actions, offering just enough detail to provide "landmarks" that assure us that no spatial realignment has occurred, while simultaneously providing spatiotemporal continuity and realism.

In the realm of sound, neither the invisible perceiver model nor the strict intelligibility/fidelity split adequately justified the increasingly heteroclite decisions to be made in the course of a film production. Simply put, the metaphors that defined the new model were the image-based principles of foreground and background. Rather than insisting upon intelligibility at the expense of naturalness or vice versa, the foreground/background principle suggests that each element of the composite be treated discretely and as "naturally" as required, but each according to different standards.

The basics of the foreground/background principle emerged in bits and pieces after 1928, along with important reconsiderations of the nature of filmic space. In particular, the theoretical role played by *perception* in defining proper sound representation came under discussion. In 1931, for example, George Lewin complains about dialogue scenes with no background noises where "we would expect various forms of background noise. . . . The dialog may be taking place in a street, and we would naturally expect to hear the characteristic street noises in the background." He adds, "It becomes necessary to put these sounds in after the picture is complete."[54] Making a practical point discussed above, he notes that technicians now prefer not to shoot on location because microphones are less selective than ears, and therefore do not ignore unwanted sounds. It is worth repeating insights. Microphones, he claims, are unable to "differentiat[e] between what we are trying to record and the background noises, for it will pick up the latter *with discouraging fidelity*."[55]

In these brief lines, Lewin articulates most of the presuppositions of the newly emergent foreground/background system. First, he redefines the role of perception in determining film sound by suggesting that, rather than unprejudiced and ideally sensitive ("*discouraging* fidelity"), it functions selectively through schemas of expectation (where "we would expect . . . noises"). It is directed and active ("capable of differentiating"), and it presupposes hierarchies of importance ("what we are trying to record"). In short, not only is the fidelity or "phonographic" model of hearing and recording inappropriate, but the newly favored model assumes that sonic reality itself is already "preheard" through conventional and often institutional expectations. The ear that simply "hears" has been replaced by the ear that listens selectively and that experiences space as a disjunctive and hierarchically ordered series of "layers," if you will. Selective listening does not *exclude* secondary sounds, however. As Lewin suggests, these backgrounds are conventional, but still essential.[56]

George Lewin explains why the foreground/background system is more flexible and better adapted to the continuity editing system developed for the image.

> The present tendency is to avoid the use of music during the shooting of the picture whenever possible, as the presence of music in the sound track hampers the editing of the picture. Without music *under the dialogue* it is possible to rearrange sequences, and make additions or omissions wherever desired when cutting the picture. This would be impossible, of course, if there were music in the track.[57]

What was true of music was true of other background sounds, too, and filmmakers began to isolate sonic foregrounds as a matter of course. So entrenched had the foreground/background model become by 1932 that the separation of sound "planes" is literalized by recording the appropriate sounds on different strips of film at different times. Picking up on their physical separation, Lewin argues against treating foreground and background as if part of a continuous and homogeneous whole, and instead advises technicians to treat them as conceptually, functionally, and even mechanically discontinuous.

Similarly, Carl Dreher complains about sudden changes in volume from shot to shot, especially evident in music, where the difficulties are compounded by the abrupt stopping and starting of melodic phrases. Musical or not, sonic backgrounds tend to be continuous and temporally unidirectional. The problems Lewin identifies with the inadvertent chopping up and reordering of incidental music caused by the editing of foreground events are a function of that very continuous and unidirectional character. While dialogue or other intermittently present sounds may stop and start, and while one image may replace another, *background* sound is a kind of "continuous melody" whose unity depends on its uninterrupted progression.[58]

The foreground/background segregation offered practical advantages, too. Lewin suggests that new strategies of film production go along with the new emphasis on foreground and background. Contrary to the tendencies of the "fidelity" approach, Lewin argues against location shooting and against recording sound effects, particularly backgrounds, "live," or along with the dialogue.

All such scenes must therefore be recorded in the studio, using an artificial set, and any *background noises which may be necessary are easily put in later by dubbing*. They can then be controlled and made to sound just as we want them to. Other examples of sound effects best put in by dubbing, which are worthy of mention, are thunder and wind noise for storm scenes, the noise of passing trains and automobiles for indoor scenes where it is desired to convey the effect that outside noises are being heard.[59]

Aside from the obvious advantage it offers for clear and intelligible dialogue, this approach also reinforces existing systems of spatial and temporal continuity. The storm and train/automobile examples are more than mere Hollywood clichés. Each offers a constant background against which foreground elements may be cut, rearranged, and situated without ever losing a sense of spatial and temporal continuity. If a passing train, for example, were recorded "live" during a dialogue scene, any editing would result in noticeable discontinuities on the sound track—a train passing during a character's impassioned plea might disappear during a reaction shot and response, only to reappear when we return to the original speaker.

It was essential that these new and emerging norms become part of the day-to-day practice of engineers—become their "gut reaction"—for the foreground/background system to take root. One manifestation of that growing institutionalization are the compensatory changes in professional roles and definitions of professional conduct that emerge in response. Several new or newly defined job titles indicate the thoroughgoing acceptance of the new norms. Levinson, for instance, argues that the "mixer" (or as we would now say, recordist) is now charged with eliminating "undesirable, even though natural, sounds," and reminds us that the "importance of having the dialog always clearly understandable goes without saying."[60] In other words, the mixer's new brief instrumentalizes specific representational norms and consequently charges him with the task of hierarchizing sonic space and effectively detaching foreground elements from their backgrounds in view of their subsequent recombination.

In a 1929 article, C. A. Tuthill similarly explains the crucial importance of the "monitor" to multitake scene construction. The best monitors, he argues, balance scientific knowledge with practical aesthetics, possessing "the brain of an engineer, but the heart of an artist." He "must have a good working knowledge of acoustics though he must not take the matter too seriously. He must be flexible enough to lend his art to all occasions, as every new set-up involves new acoustics, and other problems, differing from the preceding shot."[61] Further, he, must recognize production values both in picture and action, and devise a means for procuring a good recording without destroying the value of a spacious and beautiful set. . . . A perspective of sound in keeping with that of the camera is desirable in spite of the fact that optical illusion and lip reading somewhat correct for differences in volume.[62]

While "optical illusion" and "lip reading" are somewhat unclear, Tuthill seems to argue that differences between *expected* volume (given a particular camera placement) and *recorded* volume (for instance, a full shot with close-up sound) are overridden by other forms of continuity, including character continuities. In other words, the imperative for scale matching pales in comparison with the demand to create continuities *across* takes in the service of diegetic narrative unities.

If a technician maintains matched camera and microphone "perspectives," Tuthill implies, he will not be "flexible" enough to meet the demands of the next shot since "every new set-up involves new acoustics, and other problems, differing from the preceding shot." One or the other, individual shot or constructed series, must dominate. Tuthill decides in favor of the edited series when he examines the case of a lengthy scene that cannot satisfactorily be recorded in a single take and results in several *partly* successful performances:

We don't throw out such "takes," but hold onto them, cut them up, and splice them together. This is done in such a way that you will never know that it is not one recording; there is no fluctuation or change in quality. There again, placing of the microphones has considerable to do with that.[63]

The lack of fluctuation, that is, continuity, results from the consistent treatment of certain elements from take to take—namely, the uniform close miking of dialogue or other foreground sounds. In short, foregrounding some sounds was essential to creating continuity across takes but, as numerous writers had argued, did so at the loss of "naturalness" and "characteristic sounds." Consequently, the sonic background took on a very specific function in the production of a continuous series.

Louis R. Loeffler reiterates the priority of the edited scene over the single continuous shot, and thereby gives us a clue to the background's new function. In a discussion of the Movietone (sound-on-film) process, he argues that "a well-cut film . . . is one which embodies smoothness of continuity, proper tempo, and the intelligent use of angles and close-ups."[64] H. W. Anderson amplifies this contention by noting that "traffic noises, explosions, and whistles . . . crying babies and the pop of opening bottles, *are added* to make the film more realistic and thus more enjoyable."[65] Together, these claims suggest that two of the pressing demands—"smoothness" and realism—might be "added." But added to what? To the dialogue track. Insofar as realism and smoothness are paired, though, the issue becomes more complex and interesting. Maintaining dialogue intelligibility by close miking, technicians imply, relegates naturalness, realism, *and* smoothness to the background.[66] In fact, some technicians implicitly attributed "story continuity" and "realism" to the sound track's foreground and background respectively.[67]

Since editing rarely if ever interrupted a line of dialogue or other foreground event, and since, therefore, the background was more frequently interrupted, in practical terms "smoothness" could be produced through a carefully continuous, added sonic background. Editors developed various techniques for smoothing cuts such as "bloops" to prevent noticeable discontinuities on the sound track, usually in changes in background sound.[68] The segregation of sound-track functions offered a workable solution to the paired demands placed on sound in Hollywood—to be "realistic," but to be assimilable to the norms and standards of continuity editing.

The often-expressed fear of "losing naturalness," associated with isolating dialogue from other sounds, could be reexamined under this new description. Put in a more positive light, *adding* a continuous background allows one to *emphasize*, or *create*, a sonic continuity that parallels and supports the constructed continuities of the image. Again, Lewin emphasizes the utterly conventional nature of background, and he suggests "dubbing all such characteristic noises rather than recording them together with the action, [which] has an advantage not only as regards tone fidelity, but also from an economic standpoint." Dubbing offers great economies, he argues, since balancing foreground and background effects on the set wastes time unnecessarily: "Another important advantage of dubbing in sound effects is that stock sound tracks of these effects can be dubbed in whenever necessary. This studio [Paramount] has a record of a thunder storm which has stood in good stead in the dubbing of several pictures."[69]

By describing effects as "characteristic," he underscores the highly conventional function of sonic backgrounds. Rather than representing sounds as they might "really" be heard, technicians sought to record sounds in such a way that they would appear *recognizable*. Furthermore, by suggesting that the separate recording of background effects enhances "tone fidelity," Lewin decisively breaks with any lingering attachment to the model of the unified and firmly situated auditor.

The process Lewin describes, however, suggests a sonic landscape composed of superimposed layers, each in sharp acoustic "focus," and therefore each sufficiently legible.[70] Rather than an Italian Renaissance painting, this acoustic landscape recalls a cubist collage, or even a photomontage, consisting of discrete elements pasted together against a shared background surface. Although the cubist analogy is misleading to the extent that cubism emphasizes the disjunctions between and among the parts of the composition, it is instructive insofar as it emphasizes the extent to which an apparently "monocular" (monaural) space is actually composed of a multiplicity of sound "shots"—different points of audition—combined on a homogenizing (or at least shared) substrate or background.

K. F. Morgan describes the process of sound composition in similar terms. He, like Lewin, suggests that "real" sounds are often inappropriate, and that "characteristic" or conventional versions are more effective.[71] He also encourages using a stock sound library for generic, *legible* versions of various sounds. By emphasizing the role of the library, whose stock of sounds may be used (or "quoted") again and again in different contexts, both Lewin and Morgan implicitly argue against an approach to representation that treats each sound as a unique and unrepeatable event, and for an approach that treats sound as a signifying construct. Like Lewin, too, Morgan argues that background sound has a binding, homogenizing function. "In conclusion, the dubbing operation has been instrumental in supplying a *unity* and *finesse* as well as rhythm and *continuity* to the sound picture."[72]

Sound space, no longer theoretically defined by the passive perceptions of a securely located (and physically real) observer, is now shot through with the hierarchies of "dramatic" relationships. The principle of dramatic soundtrack relationships gets a vote of approval from no less an authority than the Academy of Motion Picture Arts and Sciences' instructor Carl Dreher, who argues that every sound man needs a thorough understanding of "story values."[73] Such understanding would, in his opinion, diminish the number of conflicts on the set, while also offering a new rationale for day-to-day technical practices. Adopting "story values" (as opposed to "fidelity values") as the practical basis for "quality control" would facilitate the integration of engineers into Hollywood by bringing their practice into accord with industry demands.

"Fidelity" as it was understood had been supplanted by a new concern with narrative effect. Thus, L. E. Clark can take the primacy of narrative for granted both when he claims that even "so simple a thing as microphone place-ment, [can] establish dramatic values through sound perspective," and when he eagerly anticipates the further development of "special circuits to distort or enhance voices."[74] Responding to Hollywood's institutional pressures, even Maxfield, ever the proponent of absolute fidelity, came to recognize the neces-sity of "a certain faking" in order to ensure an effective recording.[75]

As I noted in chapter 5, AMPAS's series of classes on motion picture sound, collected and published in 1930,[76] intervened in technical debates by deliber-ately working to establish a consensus on the sound techniques best suited to the movies. This volume culminates several years' effort at normalizing stan-dards and practices within the industry,[77] and hoped to redirect sound research and development so that they would parallel Hollywood aesthetics. By establishing a clear set of research priorities designed to solve Hollywood's sound problems (rather than broad-ranging "basic research"), technicians would address specific filmic dilemmas and solve them pragmat-ically on a case-by-case basis.

As far as the studios were concerned, basic research was simply too time-consuming and too devoted to general problems of very broad application. So while Bell investigated practical areas like "amplification" or "micro-phone sensitivity," even this kind of research was of a more general nature than suited Hollywood's sense of immediate need. Thus it is not at all sur-prising that in the 1930s many practical forms of technological innovation in sound emerged from studio shops rather than from scientific or indus-trial laboratories.[78] The institutional demands determining studio research programs justified a narrowed focus, since research was directed at specifi-cally filmic problems and functions, and the studios had little to gain by sys-tematic patent acquisition.[79] As noted, for example, studio concerns were behind the invention and innovation of the sound-proofed, telescoping microphone boom, which facilitated silent microphone adjustments during a shot and allowed recordists to keep the mike within a few feet of moving actors at all times. In fact, the demand for such a device was clearly shaped by classical *image*-editing norms and indirectly bears the traces of classical continuity assumptions—indeed, it may be said to embody them in its material form.[80]

Dreher analogizes the practices of sound space construction to special-process photography—specifically, the blue-screen process—in which the "foreground action is shot against a blank backdrop, a film background being supplied by artificial means," adding, "the blue cloth against which the fore-ground action is photographed corresponds to the silent background of dialog in sound."[81] Allowing for the substitution of any desired background, and therefore spatial and temporal continuity, the variety of emergent foreground/background practices brought sound construction into greater alignment with the norms of image construction.

Recording only the dialogue portions of the sonic whole while on the set facilitated greater control over the resulting sonic environment, and further allowed a greater interchangeability of personnel. Given the foreground/background system's reliance on generic sounds and a standard approach to dialogue recording, any well-trained technician could achieve the same result as any other. In addition, the processes of foreground/background sound segregation were cheaper and allowed a fidelity "higher than could be obtained by an original recording of the desired sound." As a bonus, libraries made previously unobtainable sounds (e.g., a meadowlark who must chirp during a love scene) readily available, and the precisely timed and recorded "stock sounds" facilitated precisely controlled and rapid production.[82]

Several years later, W. A. Mueller was to describe a device (apparently in use at Warner Bros. since 1932) that materially and functionally embodied many of the representational norms emerging in the transition period.[83] This device, which automatically controlled the balance between two sounds (known colloquially as the "up-and-downer"),[84] effectively automatized the foreground/background system of spatial construction. Mueller points to what, by 1935, seemed the obvious advantages—efficient balance between foreground and background, faster production, enhanced dialogue intelligibility—but at the same time criticizes existing techniques. Focusing on the relation between dialogue and effects, he laments the common practice of nearly eliminating the sonic background in favor of absolutely intelligible dialogue, claiming that such moments of quiet "are no longer lifelike, and the picture loses realism."[85] He further explores the foreground/background relationship by pointing out how the need for dialogue intelligibility must be balanced with a need for sonic verisimilitude—the constant presence of a background and the balance between the two constituting the basis of sonic realism.

Mueller's device renders this aesthetic nearly foolproof, allowing the dialogue track (whether it contains dialogue or not) to control the levels of all other tracks. In short, the presence of sound on the control, or foreground, track automatically reduces the volume of all other tracks without eliminating them. This device clearly assumes a radical separation and hierarchy among the components of a sonic space, one of which enunciates the narrative while the others provide verisimilitude.

Mueller's description of the up-and-downer illustrates a level of narrative understanding rarely expressed in technical articles of the period. Mueller indicates that the voice is not synonymous with the controlling dialogue track, but that important effects might function as the narrating "voice" at times.[86] Since sound editors would be responsible for the crucial step of assigning sounds to different tracks, they clearly needed to understand how classical narration operates. Rather than limiting the rationales associated with spatial construction to perceptual realism or the equally restrictive assumption of vocal primacy, Mueller's apparently offhand equation of dialogue and effects indicates an internalized awareness that at any point of a film, any particular aspect of the soundscape may dominate the narrative, and therefore assume the "foreground."

To illustrate this point, Mueller describes a crowd sequence which demands many extras, yet which also contains a fair amount of dialogue. Recognizing that it was far less expensive to shoot the crowd shots all at once and then send the extras home, he emphasizes the value of the background regulator in this situation. After dispensing with the extras, dialogue scenes could be shot in close-up whenever it was convenient, enabling the desired close miking of character's voices and without any crowd noise that might interfere. As Mueller puts it,

> By using the background regulator to control an effects loop, whenever effects or dialogue are present on the dialogue track, the background loop is suppressed. Whenever there are no effects or dialogue on the dialogue track, the background loop is recorded at full level. The result is a scene *with a smooth background*, resulting in a great improvement in the finished picture and a material saving in the cost of production.[87]

In other words, the sonic background, described again and again as the very essence of sonic realism, is now effectively reduced to a generic *loop* that repeats endlessly, in the service of the dialogue track. Again, it is clear that the need for a "smooth background," or a convincing continuity of space and time, determined the form of not only the individual takes but the hierarchies between aspects of the total sonic world.

Maurice Pivar, a film editor, perhaps best expresses the new aesthetic of sound space, which privileges the unity of the representational series over that of the individual fragment—an aesthetic toward which his job surely biased him. Reiterating, in a 1932 article for *American Cinematographer*, the problems of recording all the sounds of a scene "live," he heartily endorses the widespread use of dubbing and rerecording which would allow editors to separate tracks and manipulate them for rhetorical or discursive effect. He underscores this assertion pointing out that

> the disadvantage of recording sound effects at the same time dialogue is being recorded is two-fold; it interferes at times with the coherence of the dialogue and results in a changing volume of the various effects when the scenes which comprise the sequence are placed together. In fact, each cut is noticeable by the change in volume of this background noise.[88]

Pivar makes clear that the need to manipulate foreground and background separately is of paramount importance, but that recording both at the same time prohibits this. Enshrining the prominence of narrative above all else and the unity of the series over that of the individual take, he points out that by dubbing and recording with multiple takes, "The balance between the two may be varied at will, making the sound absolutely flexible in the hands of the dubber, and enabling him to at all times *keep the dialog intelligible* above the general noise level."[89]

As the up-and-downer's automatization of the foreground/background principle suggests, the coming of sound could hardly have threatened the primacy of narrative or narratively based aesthetic decisions—they were simply too economically successful. When describing the benefits of the balancer, Mueller claims that it "gives a *greatly improved product* with the dialog perfectly understandable at all times, the background music loud and full, and effects that are loud enough to contribute to the realism of the picture,"[90] indicating beyond any doubt that the norms of representational practice are dictated by the economic status of a particular commodity form rather than any universal standards of fidelity or accuracy. By conforming both to the economic and representational demands of the classical cinema, the up-and-downer soon became a common feature in Hollywood sound studios. What is more, this device, and the related techniques it implies, demonstrate that while the classical film produced a series of pleasures (spectacle or sensation as well as story, for example), these were, by and large, ordered hierarchically, with narrative usually determining that order. In its literal embodiment of a dual sound system, the up-and-downer seems to be saying, "spectacle, yes, but in subordination to story." (See figure 6.4.)

The fidelity-based ideas of simple hearing and lack of manipulation never die, but instead are redefined under an expanded version of the classical model. This redefinition represents the dialectical absorption of the fidelity

FIGURE 6.4
Technician adjusting the sound mix to match the visual "planes" of the image.
Important sounds, like important actors, occupy the "foreground." (© 1938 Turner
Entertainment Co. A Time Warner Company. All Rights Reserved)

mode by a new narrative mode. Whatever the practical emphasis on construction, multiplicity of points of audition, conventionality, cutting and past-
ing—in other words, on a sonic space that is (in a broad sense) constructed or
interpreted rather than simply heard—the classical film attempts to produce
the *effect* of a pure listening, that is, of a wholly prior event upon which we simply "eavesdrop." Hollywood's technicians developed a variety of techniques
that, although based on foreground/background and "constructive" principles, produced the effect of conforming to fidelity or "innocent ear" principles—engaging in what appears to be scale matching between image and
sound, or emphasizing the sonic specificity of a concrete act of listening.[91]

For instance, Carl Dreher emphasizes the necessity of maintaining proper
sound "perspective" and criticizes those who maintain the utter importance
of "close-up" sound regardless of image scale. In a move that seems to rein-
state naive fidelity principles, Dreher argues instead that image and sound
perspective should (roughly) match,[92] using as his example a scene where the

camera follows a frightened character around a house as she retreats from an "overheard" threatening conversation. As we follow her around a corner, Dreher claims,

> It would be natural for the recordist to drop the level rather sharply as the actress turns each corner, since such a variation in sound corresponds with reality. This would be taken care of automatically if the microphone dollies with the camera, but if this effect is not secured in the original recording, it may be simulated in re-recording.

One cannot, it seems, contradict image scale, willy-nilly. Bolstering his argument, he describes a scene where a character addresses a crowd from a high platform, but where an uninformed editor had maintained close-up sound uniformly, depite a camera position on the ground. Here, the effect was ludicrous, he argues, and destroyed the sense of massive scale produced by the set.[93] However, despite its apparently "natural" foundation, its basis in real perception, and Dreher's implicit relinking of mechanical eye and ear, this particular use of perspective is no less a convention than is the foreground/background principle. In Dreher's first example, the diminishing sound is clearly meant to be understood as heard from the frightened woman's point of audition (POA). Distant miking in the second example likewise represents a less individual POA from within the crowd. In each case a literal sense of spatial dimensions is *narratively* rather than perceptually justified.[94] In other words, these instances of sonic perspective do not follow from any consistent application of "realistic" representational norms, but from the shifting demands of an omniscient narration. In these cases, the classical film employs a noticeable use of "perspective" because the narration posits a listening agency *within* the space of the diegesis. It thereby literalizes the surrogate perceiver, and so, contrary to its apparent perceptual basis, sound space is still determined by narrational demands, like the foreground/background system, even if the hierarchy is less obvious.

Such uses of sound perspective are generally justified by a shift from a less restricted to a more restricted level of narration.[95] The most common form of restriction entails presenting sound as if *heard* by an embodied listener. POA sound is not entirely restricted to the hearing of characters within a fictional diegesis, however, and here we see how the lessons of the Vitaphone shorts can be absorbed and refashioned by the Hollywood style. Many of the early Vitaphone shorts, particularly the musical ones, offer film spectators a single, unchanging point of audition that allows them to hear the full quality of the acoustic space—like a POA take. In this case, however, a fictional character's hearing within a fictional world is not obviously at issue. Still, a similar logic obtains. In these cases where the diegesis attempts to mimic live theater, the *spectator* is the "character" whose presence within the space is represented by the use of perspective and who is, implicitly, *within* the "diegetic" space.

As I noted earlier, the presentational style, frontality of performance, and mode of address typical of the shorts implies the presence of an audience within the space of representation (which is here the diegesis as well). Of course, as Charles Wolfe points out, the "direct address" to the film audience requires the fictionalizing gesture of ignoring the actual presence of the camera, mike, and crew.[96] Although representation is here figured as simple documentation, with performers acknowledging the real presence of an imagined future audience, the process of representation is every bit as narrational as the production of a fictional diegesis. In this case, however, the listening "character" constructed by the text is meant to be us, the spectators in the movie theater.

In this respect, we are not far at all from the practices of the classical narrative, but simply offered only one, "restricted" narration of the event. Doubtless, this effect is rarely *experienced* as a form of restricted narration or focalization because we are explicitly addressed in the text *as* hearers, yet the process by which the space of reception is joined to the space of representation (while different in *effect*) is essentially the same in form as that which binds us to the diegetic space by means of a character-focalized narration. We see, however, that such techniques are not *necessarily* part of an alternative mode, but may take their place within a larger, more flexible, and variegated approach to classical sonic construction.

Dreher himself points to the ultimate logic underlying the construction of sonic space via two brief discussions that explicitly introduce a foreground/background metaphorics as the justification for particular spatial configurations. After asserting the utter necessity of "naturalness," or "acoustic fidelity to the original rendition," he argues that

> the sound should in general follow the action so that the loudness will decrease with increasing distance in the picture, but . . . even at its minimum, the sound must be loud enough to be clearly understood if the action of the play requires it. (There are, of course, instances when a mere murmur of voices or certain intelligible lines standing out above unintelligible material are all that is required.)[97]

Once again, even in a defense of perceptual realism, yet another technician identifies narrative as the motivating source of images and sounds. Rather than records of perceptually specific events, sounds here are signifiers of an absent, but inferred, narrative.

Against the repeatedly expressed ideals of the period's leading sound technicians and the allure of an aesthetic based on a perfectly duplicative mode of representation, the demands of the industry within which they worked reordered the norms, practices, and even the very assumptions that structured their representational practice. Despite some momentary challenges to the classical norms of spatial construction embodied in the Vitaphone shorts and the early part-talkies (and despite the persistence of these different challenges in alternative modes like the documentary and the avant garde), the overwhelming pressure for profitability and consistency of both mode of production and product mandated a system of practices which contradicted the most often explicitly voiced representational goals. Rather than simply capitulate to Hollywood's demands, however, these technicians struggled for control over their own decisions. Nevertheless, through a reconceptualization of their own role, and of the aesthetics appropriate to their profession, they ultimately assimilated themselves to a new representational culture, made its norms their own, and over the years came to regard them, as they had the norm of absolute fidelity, as simply natural.

CONCLUSION

The interplay of senses, technology, and aesthetics outlined in the preceding pages haunts us still, shaping our reactions to and understandings of current representational technologies. The scenes we witnessed in the nineteenth century found experimenters explicitly modeling sound devices (for both producing and recording) on the human body in the belief that there was no other logical, or indeed appropriate, way to proceed. Bolstered by a belief in the perfection of God's design of the human voice and sensorium, mechanical attempts to emulate the human became the royal road to scientific knowledge. However obvious, the ideal of human simulation was not the only or the inevitable way to think about sound technology. The other major impulse driving the investigation of sound and the development of sound (recording) devices was to understand the process as a form of writing.

Representation-as-writing found its roots both in the graphical analysis of subtle vibratory movements begun by Robert Willis and E. F. F. Chladni and in the contemporaneous movement to create a nonarbitrary, analogical, and universal form of writing. The *phonautographe* and its many scriptural offspring, including Edison's phonograph, recorded vibrations in the air, impressing them on foil, lampblack, or wax, in what some described as "nature's own script." Even these writing machines were not immune to the drive toward simulation, however, and many were, in fact, constructed from the dissected ears of dogs, cats, and even humans. Embedded in design, purpose, and result, simulation and inscription circumscribed the imagination of these technologies, for better or worse. Consequently, when these new devices surpassed or confounded expectations about what a sense could do, their novelty was understood through a combination of perceptual and scriptural tropes.

Sound technology's dual roots in human simulation and in forms of inscription decisively shaped the earliest encounters with new sound media. If nothing else, these encounters wed the two tropes in an indissoluble pair that allowed people to understand inscription as a form of (mechanical) sensation, and sensation as a form of writing. Indeed, recognizing that both tendencies were interrelated set the stage for the reception of the technical media more generally. Since emerging sound technologies were understood as literally as possible both *as* senses and *as* forms of writing, whatever dilemmas of a practical or epistemological sort they raised were addressed through the concepts of sensation/perception and of writing. Reciprocally, commentators began to think of the human senses themselves as mechanical. The idea that sensation could incorporate a species of writing—a kind of prosthetic memory—suffused the vernacular imagination, and expressions like "photographic memory" became commonplace. By the same token, these extraordinary new sensory capabilities were balanced by the recognition that such "writings" might be quoted and/or taken out of context. Indeed, a recorded voice could long outlive its speaker or contradict him in his own presence. Senses that could store experience could also render it unstable, mobile, and public. The entire process served to question the limits of and nature of human experience.

From one direction, then, the human served as a model and ideal for the mechanical. From the opposite direction, the mechanical began to threaten the very idea of the "human" upon which it was based. In a subtler corollary, the mechanical eventually emerged as a model for the human to emulate. In a similar fashion, writing served first as a model for machines, then as a way of marking the nonhuman aspects of mechanical sensation—their absence, lack of authenticity, and such—and finally as a way of reimagining the human and the superhuman. In essence, tropes of simulation humanized the processes of mechanical inscription, while those of writing estranged it. Together, however, they ultimately reaffirmed the centrality of the human subject to a series of processes that seemed, at times, utterly to exclude it.

This tropological pairing persisted throughout the period I examine, confronting the same kinds of issues it did in the nineteenth century, but not in a simple or straightforward way. Simulation, for its part, remains to this day a fairly unchanged ideal. Perceptual and representational technologies (like

3-D images and stereo sounds) still take the human body and human perceptual experience as a sine qua non. Inscription, in contrast, has taken a much more circuitous path. From its earliest appearance as a literal account of a mechanical process, "writing" began to take on different guises: an "absent" substitute for speech, a paradigm for visual and acoustic "legibility," and a metaphor for all of those aspects of a represented phenomenon that are *constructed* and *signifying* rather than simply "present" in some putative "original" that it copies. In our day, it hovers around the idea of the digital, whose discontinuous, nonanalogical, and exactly repeatable inscriptions form the basis of a current redefinition of human capacity. Today, just as one rarely hears an exquisite memory described as "photographic," one can hardly avoid similar analogies derived from the realm of the computer. Memory (even human memory) is now conventionally understood to be digital, randomly accessible, and stored in discrete packets of data. The photograph, by comparison, seems a quaint and altogether too sensual medium to describe memory.

So, while scenarios of technological encounter are indeed repeated over historical time, I don't mean to say that the pair simulation/inscription are not historical in nature. They are not essential categories of the technologies, but rather, historically specific and historically developed approaches to and responses to the project of developing sensory technologies. To the degree that they are products of a particular Enlightenment-era project to investigate and understand the senses, they are resolutely historical, and even antique. To the extent that the issues motivating their emergence are not yet settled, they are contemporary. Whatever repetitions occur, they occur because the basic issues of the relationship between perception, representation, and experience are still very much in question.

Thus, it should be no surprise that the coming of sound to Hollywood during the 1926–1934 period was both a unique moment in the history of both the film and recording industries, and just as surely a familiar and even predictable one. As technicians came to grips with the possibilities and problems specific to the medium of recorded sound, they revisited terrain previously explored in the 1880s–1910s, and 1920s–1930s when earlier generations grappled with the meaning of the phonograph and its appropriate forms. Confronted with a medium that seemed inherently "realistic" and a set of devices that seemed inherently "sensory," sound researchers, engineers, and studio personnel responded by developing not only an aesthetic of perceptual simulation but a set of related research priorities as well. As was the case at earlier moments, the model of representation as sensory duplication quickly ran afoul of entrenched aesthetic norms and institutional demands. Whether spurred by the music industry's need to preserve the sound of quieter instruments amid their noisier collaborators, or Hollywood's need to construct artificial visual and sonic continua, the premise of simulation proved inadequate to a variety of basic recording tasks. Institutionally mandated practices that contradicted naturalistic assumptions of perceptual surrogacy required technicians to rethink previously unquestioned standards, and move further and further from representational models that took the human body and the human witness as its principle of unity and its ideal.

Still, the tendency to understand technical media as surrogate senses and thereby to assume that one's aesthetic decisions were determined by a dogged faithfulness to the virtual body of a sensory witness persisted across the decades. Whether 3-D visual processes or stereo sound, new representational technologies typically passed through a stage where they were understood quite literally as sense organs. Said organs were more sensitive than human organs, and capable of storing sensory appearance, but emergent aesthetics invariably translated the notion of perceptual simulation into irrefutable and rigid principles of both local and global formal construction. When the trope of simulation shifts from a strategy to account for unexpected deviations from human experience to an aesthetic precept, it becomes most vulnerable. The conflict between an irresistible perceptual aesthetic and the immovable demands of an industry throws the limits of any single technological model into sharp relief. The case of 3-D film makes this point clear.

When rediscovered and given new life through technological innovation in the mid-1950s, 3-D film processes seemed ready to transform the entire industry. Recognized as a potential weapon to combat the encroachments of television, 3-D was at the outset almost universally described as a "more realistic" form of filmmaking. Taking its cues from the physical organization of the body (human eyes and 3-D camera lenses both set 2.5 inches apart), 3-D systems tried to outstrip television and conventional cinema by offering a more compelling sensory illusion. According to one typical account, "the ideal stereogram should be a 'real-life' view as seen through a window." Amplifying his claim, the writer argues that 3-D films culminate the search for a transcendent realism in the visual arts: "No graphic means, besides the stereogram, can substitute for the re-creation of the 'real' in a still life, and in stereo movies realism reaches the ultimate, for they can include movement, color, and action, as well as depth."

This essential property translates into aesthetic rules quite naturally, since "the apparent depth of the stereoscopic view should be the same as the real depth of the scene."[1] Of course, no sooner are these universals established than they are contradicted. The "real-life view" seen in the stereogram, for instance, should have no objects in the foreground, we are told, lest they confuse the 3-D effect. Even more troublesome, it was learned, was the complications 3-D brought to classical editing norms. As Raymond Spottiswoode argues at great length, 3-D images introduce an extra dimension to spatial construction, and cutting in familiar continuity fashion may cause newly "dimensioned" objects suddenly to leap in and out of the screen during the most innocuous match-on-action edits.[2]

While I do not want to dispute the premise that 3-D images may very well offer a more vivid sense of perceptual immediacy and immersion, the example of 3-D should make it clear why we need actively to resist the lure of the simulation argument and the pull to generalize the ideal of perceptual simulation to the point of aesthetic rule. In brief, this pull encourages us to mistake perceptual space for narrative space, or, more generally, it misleads us (by way of its apparent obviousness) into believing that the processes of representation are, in fact, simply the processes of perception. So strong is this temptation that this very same confusion emerges at nearly the same historical moment with regard to stereo film sound, and in almost identical fashion in the mid-1980s with the advent of stereo TV sound. In every case, long-recognized (and far from literally human) rules of spatial construction are momentarily reassimilated to the human body and to a wholly nonmechanical way of apprehending the world. Only later, and under the pressure of practical exigency, do the two once again separate.

This dilemma, in effect, is both the crux of our apparent compulsion to repeat nineteenth-century encounters in our engagements with 21st-century media, and the link between technology and the project of modernity. As I note repeatedly throughout the book, the impulse toward understanding representational or perceptual technologies as simulations of human capacities is balanced by the equally powerful recognition of the truly inhuman tendencies they just as surely embody. The historically repeated gesture of rendering the mechanical human responds to the basic dilemma produced by the emergence of the technical media, and of the possibility of technological inscription specifically. To wit: when machines may see and hear phenomena and depict them with an ease unknown to mortal artists, what becomes of established representational epistemologies?

The photograph, the phonograph, and the cinema each asks us to examine the same basic questions. "What is the nature of technologically mediated sensory experience, and how does it relate to technologically produced forms of representation?" Put more bluntly, the devices ask us to consider the principle of representational unity of the technologically produced sign. Is the phonograph truly a surrogate ear? If it is more sensitive, must we automatically trust its recordings or does its lack of consciousness allow us to assert our own power to judge? Under what conditions can we accept the "vision" of a camera as truthful and accurate? How does a snapshot relate to my unmediated encounter with the same visual event?

Much of the point of this book has been to argue that the answers to such questions are given neither by the nature of the media under discussion nor by the nature of specific devices. These are not, when all is said and done, "ontological" questions or essential "properties." Instead, they are pragmatic, historical, and contingent forms of knowledge produced in response to concrete and objective material possibilities, but also in response to emerging discourses, existing practices, and established institutions. Any one of these dimensions may decisively alter the answer we give to any epistemological question we pose regarding our sensory relationship toward the world, as it is refracted through the media of technological representation.

NOTES

Introduction—Discourse / Device / Practice / Institution

1. Dana Brand, "From the *flâneur* to the Detective: Interpreting the City of Poe," in Tony Bennett, ed., *Popular Fiction: Technology, Ideology, Production, Reading* (London and New York: Routledge, 1990), 220–37; see also Walter Benjamin, *Charles Baudelaire: A Lyric Poet in the Era of High Capitalism*, trans. Harry Zohn (London: New Left Books, 1973), 48. The inspiration for my interpretation is Robert Ray, "Snapshots: The Beginnings of Photography," in Dudley Andrew, ed., *The Image in Dispute: Art and Cinema in the Age of Photography* (Austin: University of Texas Press, 1997), 293–308.

2. Richard Sieburth, "Same Difference: The French *Physiologies*, 1840–1842," in Norman F. Cantor, ed., *Notebooks in Cultural Analysis* (Durham, N.C.: Duke University Press, 1984), 163–200.

3. Allan Sekula's "The Body and the Archive," *October* 39 (1986): 3–64, esp. 18–20, offers a compelling account of the logic informing each project.

4. Siegfried Kracauer, "Photography," *The Mass Ornament: Weimar Essays*, trans. and ed. Thomas Y. Levin (Cambridge, Mass., and London: Harvard University Press, 1997), 47–63. See Benjamin's related comments on the "optical unconscious" in "A Short History of Photography" and "The Work of Art in the Age of Mechanical Reproduction," *Illuminations*, trans. Harry Zohn (New York: Schocken Books, 1969), 217–52.

5. Miriam Hansen, "America, Paris, the Alps: Kracauer (and Benjamin) on Cinema and Modernity," in Leo Charney and Vanessa Schwartz, eds., *Cinema and the Invention of Modern Life* (Berkeley, Los Angeles, and London: University of California Press, 1995), 362–402.

6. Jean-Louis Baudry, "Ideological Effects of the Basic Cinematic Apparatus," in Philip Rosen, ed., *Narrative, Apparatus, Ideology: A Film Theory Reader* (New York: Columbia University Press, 1985), 289.

7. Baudry, "Ideological Effects," 286–90.

8. Julia Kristeva, *Sémeotikè: Récherches pour une Sémanalyse* (Paris: Seuil, 1969), 13, cited in Jean-Louis Comolli, "Technique and Ideology," in Rosen, ed., *Narrative, Apparatus, Ideology*, 424.

9. Comolli, "Technique and Ideology," 435–38.

10. The reconceptualization of a single device—the radio—from an instrument for point-to-point communication to a medium for broadcast exemplifies this fluidity.

1. Inscriptions and Simulations: The Imagination of Technology

1. Thomas Edison, Caveat 110, October 8, 1888, filed October 17, 1888, patent records, NjWOE, cited in Charles Musser, *The Emergence of Cinema: The American Screen to 1907*, vol. 1 of *History of the American Cinema*, ed. Charles Harpole (Berkeley, Los Angeles, London: University of California Press, 1990), 64. Gordon Hendricks transcribes the first four motion picture caveats in his *The Edison Motion Picture Myth* (New York: Arno, 1972), 158–63.

2. F. Nadar, "Daguerréotype acoustique," *Le Musée franco-anglais* (1856) and *Les Mémoires du géant* (Paris, 1864). Both are cited in Jacques Perriault, *Mémoires de l'ombre at du son: un archéologie de l'audio-visuel* (Paris: Flammarion, 1981), 133–34.

3. So completely did the nineteenth-century imagination seek the transformation of daily life through technology that photography, phonography, and telegraphy all came to inhabit the discourse and methods of spiritualism. Spirits at séances, for example, only began to communicate by "rapping" after the invention of the telegraph, turning later to photography both as a means of communication and as metaphorical understanding of spiritual manifestation. One can be sure that the phonograph and radio have their analogs here as well. For a penetrating analysis of spirit photography, see Tom Gunning, "Phantom Images and Modern Manifestations: Spirit Photography, Magic Theater, Trick Films, and Photography's Uncanny," in Patrice Petro, ed., *Fugitive Images: From Photography to Video* (Bloomington: Indiana University Press, 1995), 42–71.

4. Thomas Edison, "The Phonograph and Its Future," *North American Review* 126 (May-June 1878): 527–36 (527). See, ten years later, his "The Perfected Phonograph," *North American Review* 146 (May 1888): 641–50.

5. On the theme of technological utopianism generally, see Walter Benjamin's remarks on Fourier, "Fourier; Or, the Arcades," *Charles Baudelaire: A Lyric Poet in the Era of High Capitalism*, trans. Harry Zohn (London: Verso, 1983), esp. 159–60. For commentary on and documentations of technological utopianism in the United States, see Howard P. Segal, *Technological Utopianism in American Culture* (Chicago: University of Chicago Press, 1985); Leo Marx, *The Machine in the Garden: Technology and the*

Pastoral Ideal in America (New York: Oxford University Press, 1964); Henry Nash Smith, ed., *Popular Culture and Industrialism, 1865–1890* (Garden City, N.Y.: Anchor Books, 1967), esp. part 1, "Nationalism, Progress, Technology"; James Richardson, "Travelling by Telegraph," *Scribner's Monthly* 4 (1872): 1–23, reprinted in Alan Trachtenberg, ed., *Democratic Vistas, 1860–1880* (New York: George Braziller, 1970); and Lewis Mumford, *Technics and Civilization* (New York: Harcourt Brace, 1934), esp. chapters 4 ("The Paleotechnic Phase"), 5 ("The Neotechnic Phase"), and 6 ("Compensations and Reversions").

6. See, among others, Roland Gelatt, *The Fabulous Phonograph, 1877–1977*, 2d rev. ed. (New York: Macmillan, 1977), 23–24; Oliver Read and Walter L. Welch, *From Tin Foil to Stereo: Evolution of the Phonograph* (Indianapolis, Kansas City, New York: Bobbs-Merrill, 1976), 6, 10, 119–21; Peter Ford, "History of Sound Recording: The Age of Empiricism (1877–1924)," *Recorded Sound* 1.7 (Summer 1962): 221–22; Harry Geduld, *The Birth of the Talkies* (Bloomington: Indiana University Press, 1975), 3–6.

7. English in the original. Villiers de l'Isle-Adam, *Oeuvres complètes*, vol. 1, ed. Alan Raitt et al. (Paris: Gallimard, 1986), 769. My information about composition and publication dates comes from this edition, as well as Alan Raitt, *The Life of Villiers de l'Isle-Adam* (New York: Oxford University Press, 1981). While published in 1889, *L'Eve future* was completed in 1886 (see *Oeuvres complètes* 1:1429–59).

8. There is, however, some final uncertainty about whether Hadaly is, in fact, an automaton. At the end, she insists she is not.

9. Among the most provocative readings of Villiers's work is Annette Michelson's "On the Eve of the Future: The Reasonable Facsimile and the Philosophical Toy," *October* 29 (Summer 1984): 3–20. Michelson links Villiers's novel to Baudelaire's ongoing critique of modernity, capitalism, and rationalization, and to the emergence of the cinema. On Villiers, see also Friedrich Kittler's brief comments in *Discourse Networks, 1800/1900* (Stanford, Calif.: Stanford University Press, 1990), 230–31.

10. Edison, "The Phonograph and Its Future," 528–29.

11. Edison, "The Perfected Phonograph," 650. For information on photographic surveillance, especially on the "right to privacy for thoughts, emotions, sensations," including "facial expressions," see Samuel D. Warren and Louis Brandeis, "The Right to Privacy," *Harvard Law Review* 4 (December 1890): 193–220; E. L. Godkin, "The Rights of the Citizen: To His Reputation," *Scribner's Magazine* (July 1890). Robert Mensel has argued that citizens found themselves "responsible for the visible evidence of their feeling, there was no escape from embarrassment, and if that moment were captured in a photograph, embarrassment could be perpetual." "Kodakers Lying in Wait: Amateur Photography and the Right of Privacy in New York, 1885–1915," *American Quarterly* 43.1 (March 1991): 31.

12. Joel Snyder, "Picturing Vision," *Critical Inquiry* 6 (1980): 499–526.

13. Carolyn Marvin, *When Old Technologies Were New* (Oxford and New York: Oxford University Press, 1988), 222–31.

14. See, for example, Jacques-Louis Mandé Daguerre, "Daguerreotype," in Alan Trachtenberg, ed., *Classic Essays on Photography* (New Haven: Leete's Island Books,

1980), 11; *Porter v. Buckley*, 147 F. 140 (3d Cir. 1906); J. H. Fitzgibbon, "Daguerreotyping," *Western Journal and Civilian* (1851); William Henry Fox Talbot, *The Pencil of Nature* (London, 1844); Nathaniel Willis, "The Pencil of Nature," *The Corsair* (April 13, 1839), 71; Anon., "New Discovery in the Fine Arts," *New Yorker* (April 20, 1839); Anon., "New Discovery in the Fine Arts," *New Yorker* (April 13, 1839).

15. This becomes a prominent motif in early film narratives as well. See, for example, *The Story the Biograph Told, Getting Evidence*, and *Traffic in Souls*.

16. On the notion of "autographic" images and related ideas, see Thomas Y. Levin, "For the Record: Adorno on Music in the Age of Its Technological Reproducibility," *October* 55 (Winter 1991): 23–47; and Patrick Maynard, "Talbot's Technologies: Photographic Depiction, Detection, and Reproduction," *Journal of Aesthetics and Art Criticism* 47.3 (Summer 1989): 263–76. On autography, more broadly considered, see François Dagognet, *Étienne-Jules Marey: A Passion for the Trace*, trans. Robert Galena with Jeanine Herman (New York: Zone Books, 1992), esp. 39, 43, and passim.

17. For example, see Marey's comments in "The Natural History of Organized Bodies," *Annual Report of the Board of Regents of the Smithsonian Institution for 1867* (Washington, D.C.: Smithsonian Institution Press, 1868), 286, where he notes that in the study of the body's processes, the unaided senses are insufficient, being "baffled alike by objects too small or too large, too near or too remote, as well as by motions too slow or too rapid." The inadequacy of the senses is a persistent theme in Dagognet's study, as a way of distinguishing Marey's approach from that of his contemporaries (see *Étienne-Jules Marey*, esp. 20, 23, 30, 46, 52, 62).

18. Such views were characteristic of responses to both Marey's and Muybridge's photographs. They were more also typical of responses to instantaneous photos, where previously invisible phenomena asserted themselves aggressively. Similar descriptions follow the emergence of the phonograph, which is commended for its capacity to record subsonic and supersonic vibrations, and then render them audible by adjusting the speed of the playback. Edison likens the phonograph to the instantaneous photograph, since each makes available to the senses nuances that escape ordinary perception (see "The Perfected Phonograph," 648–49). Additionally, he attributed his success at developing acoustic instruments, especially the phonograph, to his desire to overcome his own deafness. Cf. Perriault, *Mémoires*, 152.

19. Alfred M. Mayer, "On Edison's Talking-Machine," *Popular Science Monthly* 12 (March 1878): 719. Mayer was a prominent physicist known by his contemporaries as a specialist in acoustics. In a gesture of historical forgetting typical of both the history of science and the history of culture, which have denigrated the acoustic, he is remembered almost entirely for his "floating magnet demonstration," which provided an early analog for atomic structure. Also see his *Sound: A Series of Simple, Entertaining, and Inexpensive Experiments in the Phenomena of Sound for the use of Students of Every Age* (London: Macmillan, 1891).

20. Richard Altick, *The Shows of London: A Panoramic History of Exhibitions, 1600–1862* (Cambridge, Mass., and London: Belknap Press, 1978), 353–56. The Euphonia played to amazed but scarce crowds at the famous Egyptian Hall through-

out the summer that year. The Duke of Wellington himself, after determining its authenticity, gave it an endorsement. The *Quincy (Illinois) Herald* for February 9, 1844, indicates that the device had appeared in the United States at least two years before its London run at the Hall. "A German, named Faber, . . . in New York, has invented and brought to perfection a talking machine. It is played on by keys like a piano, and can be made to say anything, in any language that the inventor desires." I have been unable to locate other references to this exhibition.

21. Like previous machines, and in full accordance with the tradition of physiological imitation, Euphonia was given a head, face, and body, and dressed as a person. Some reports suggest that, like contemporary automata of other sorts, it was dressed as a Turk. The only illustration I know of (reprinted in Altick, *The Shows of London*, 355; figure 1.2, this volume) shows a mannequin of indeterminate ethnicity or nationality, without the usual trappings of "the Turk." Compare pictures of Maelzel's infamous chess-playing "Automaton" in Charles Michael Carroll, *The Great Chess Automaton* (New York: Dover, 1975).

22. See Altick, *The Shows of London*, 356. The husband of Faber's niece apparently toured with the Euphonia under the name "Professor Faber." For information on Barnum's relationship with Euphonia, see A. H. Saxon, *P. T. Barnum: The Legend and the Man* (New York: Columbia University Press, 1989), 150–51.

23. Mayer, in fact, mentions the term "phonograph" at the outset of his essay ("On Edison's Talking-Machine"), only to translate it "adopting the Indian idiom" as "The Sound-Writer *who talks,*" introducing a speaker where none exists etymologically.

24. Deborah Jean Warner, "What Is a Scientific Instrument, When Did It Become One, and Why?" *British Journal for the History of Science* 23 (1990): 83–93.

25. Mayer, "On Edison's Talking-Machine," 719.

26. This is how one rather sympathetic memoir describes Faber's Euphonia. John Hollingshead, *My Lifetime*, vol. 1 (London, 1895), 67–69, cited in Altick, *The Shows of London*, 355.

27. Alexander Graham Bell, "Prehistoric Telephone Days," *National Geographic* 41.3 (March 1922): 228. On this story, see also Robert V. Bruce, *Bell: Alexander Graham Bell and the Conquest of Solitude* (Boston and Toronto: Little Brown, 1973), 35–37; William T. Jeans, *Lives of the Electricians: Professors Tyndall, Wheatstone, and Morse* (London, 1887), 111–16, 117–18.

28. Isaac Pitman, *Phonography; Or, Writing by Sound: Being a Natural Method of Writing, Applicable to all Languages, and a Complete System of Shorthand* (London: S. Bagster, 1840). As Thomas Y. Levin notes in an important essay ("For the Record: Adorno on Music in the Age of Its Technological Reproducibility," 36), phonography was vigorously promoted by workers' groups as a way to make writing more accessible. Marey, on the other hand, was driven by the need to produce a direct, continuous script. As Dagognet puts it, "He directly questioned not only our senses, which deceive us, but also languages that lead us astray. . . . The 'trace' in contrast, was to be considered nature's own expression, without screen, echo or interference: it was

faithful, clear and above all, universal" (*Étienne-Jules Marey: A Passion for the Trace*, 62–63).

29. See, among other pieces, Edouard-Léon Scott [de Martinville], *La problème de la parole s'écrivant elle-même: La France, l'Amerique* (Paris, 1878). The earliest publications on the device date from 1857, when he presented his method to the Académie des Sciences.

30. Dagognet notes that Marey, too, considered his own method "exegetical"— that is, as a system not only of writing but of reading (see *Étienne-Jules Marey*, 51–52, 62, and passim).

31. Dagognet, *Étienne-Jules Marey*, 63.

32. Bell, "Prehistoric Telephone Days," 228.

33. For a useful account of early phonograph concerts and several examples of early programs, see Charles Musser and Carol Nelson, *High-Class Moving Pictures: Lyman H. Howe and the Forgotten Era of Traveling Exhibition, 1880–1920* (Princeton: Princeton University Press, 1991), ch. 3.

34. Bell, "Prehistoric Telephone Days," 233–34.

35. Kempelen appears, most memorably, as a ghostly presence at the beginning of Benjamin's "Theses on the Philosophy of History," where a "hunchback" hidden in the automata and hidden by "mirrors," secretly manipulates the chess game. The hunchback (rather than Kempelen) is the first image of historical materialism. Benjamin most likely encountered the chess-playing automaton from Poe, through Baudelaire, since he repeats two of Poe's errors—namely, the supposition that the device required a man of unusually small size, and that he was hidden by mirrors. Neither was true. These inaccuracies are irrelevant to Benjamin's point, although if corrected, the image might lose a certain vividness.

36. Bell, "Prehistoric Telephone Days," 235.

37. Edouard-Léon Scott [de Martinville], "Inscription automatique des sons de l'air au moyen d'une oreille artificielle," *Comptes Rendus* 53 (1861): 108. See, as well, Clarence J. Blake, "The Use of Membrana Tympani as Phonautograph and Logograph," *Archives of Ophthalmology and Otology* 5 (1876): 108–113. Bruce describes the importance of Blake's contributions of dissected ears to Bell's researches (see *Bell*, 112–21).

38. In addition to Kempelen, the Abbé Mical and Erasmus Darwin constructed talking machines based upon the principle of exact physiological imitation. Félix Savart used as evidence of his theory of vowel production (which explained the phenomena morphologically) several plaster casts of human cadavers. Charles Cagniard de la Tour continued this line of research into the 1830s when he made a number of rubber models of the human larynx in an attempt to produce vowels.

39. Dayton Clarence Miller, *Anecdotal History of the Science of Sound: To the Beginning of the 20th Century* (New York: Macmillan, 1935); James Loudon, "Century of Progress in Acoustics," *Science* (December 27, 1901): 987–995; Frederick Vinton Hunt, *Origins in Acoustics: The Science of Sound from Antiquity to the Age of Newton* (New Haven and London: Yale University Press, 1978), 43–81.

40. David Brewster and Charles Wheatstone, *Brewster and Wheatstone on Vision*, ed. Nicholas J. Wade (London and New York: Academic Press, 1983); Charles Wheatstone, *The Scientific Papers of Sir Charles Wheatstone* (London: Taylor and Francis, 1879). In the latter, see especially "Description of the Kaleidophone, or Phonic Kaleidoscope; A New Philosophical Toy, for the Illustration of several Interesting and Amusing Acoustical Phenomena" (21–29), and "On the various Attempts which have been made to imitate Human Speech by Mechanical Means," *Report of the British Association—Transactions of the Sections* (1835) (14).

41. David Brewster, *Letters on Natural Magic addressed to Sir Walter Scott, Bart.* (London: John Murray, 1832), esp. chs. 1 and 7.

42. See Gerard L'Estrange Turner, "Scientific Toys," *Scientific Instruments and Experimental Philosophy, 1550–1850* (Brookfield, Vt.: Variorum, 1990): 377–98, for a useful account of the role such books played in the dissemination of science. For related insights see Barbara Stafford, *Artful Science* (Cambridge: MIT Press, 1994), chs. 1 and 2. Pepper was also famous for the illusion known as "Pepper's Ghost," a magic lantern projection that enjoyed a great success at the London Royal Polytechnic Society in the 1860s when he was its director. See Erik Barnouw, *The Magician and the Cinema* (New York and London: Oxford University Press, 1981).

43. On the role of public demonstration, see Stafford, *Artful Science*, passim; Simon Schaffer, "Natural Philosophy and Public Spectacle in the Eighteenth Century," *History of Science* 21 (1983): 1–43; Stephen Shapin and Simon Schaffer, *The Leviathan and the Air-pump* (Princeton: Princeton University Press, 1985); and David Gooding, Trevor Pinch, Simon Schaffer, eds., *The Uses of Experiment: Studies in the Natural Sciences* (Cambridge: Cambridge University Press, 1989).

44. This neglect is prominently a part of the historiography of science as well. Despite its status as the premier experimental science of the nineteenth century, acoustics barely figures in histories of the period. For valuable analyses of the neglect of both acoustics and experimental science more generally, see Dayton Clarence Miller, *Anecdotal History of the Science of Sound*, and V. Carlton Maley, *The Theory of Beats and Combination Tones, 1700–1863* (New York: Garland, 1990), among others.

45. See, for example, Charles Wheatstone, *Scientific Papers*, especially "Description of the Kaleidophone, or Phonic Kaleidoscope" and "On the Transmission of Musical Sounds through Solid Linear Conductors, and on their Subsequent Reciprocation," 47–63.

46. Wheatstone, "On Reed Organ Pipes, Speaking Machines, etc.," in *Scientific Papers*, 348–67.

47. Kempelen, a member of Hungarian court society, distinguished himself as court counselor of the exchequer, a member of the court chancery, designer of the hydraulic system for the fountains at Schönbrunn Palace, and supervisor of the salt industry. The best account of his career is Carroll's *The Great Chess Automaton*, 1–3. Poe's essay, "Maelzel's Chess Player," *Southern Literary Messenger* 2 (1836): 318–26, makes a great many errors, primarily because he repeats the mistakes of a previous writer from whom he plagiarized the bulk of his piece (see Carroll, ibid., 82–85).

48. Willis accomplished this by proposing that the study of vowels not be linked directly to an analysis of human morphology but to the more general study of vibrations, pure and simple. "The vowels are mere affectations of sound, which are not at all beyond the reach of human imitation in many ways, and not inseparably connected with the human organs, although they are most perfectly produced by them." In an essay on Wolfgang von Kempelen's talking machine, Homer Dudley and T. H. Tarnoczy note that Willis showed that the different shapes of Kempelen's chambers were not necessary to achieve different vowels, just different resonances (see "The Speaking Machine of Wolfgang von Kempelen," *Journal of the Acoustical Society of America* 22.2 [March 1950]: 157).

49. See Marey's basic treatise, *La méthode graphique dans les sciences experimentales*, 2d ed. (Paris: Masson, 1885). As Dagognet notes, advances in acoustical research, especially in the area of graphic representation, were absolutely crucial to the development of Marey's method (see *Étienne-Jules Marey*, 37–44).

50. Koenig perfected the manometric flame and made crucial improvements to Scott's *phonautographe*—the premier inscriptional instruments of their day—and his tuning forks were the world's standard.

51. H. M. Boettingor, *The Telephone Book: Bell, Watson, Vail, and American Life, 1876–1976* (Croton-on-Hudson, N.Y.: Riverwood, 1977), 51–52.

52. Unlike the resonance theory, which argued that specific ear structures were "tuned" to different frequencies, and thereby could "analyze" complex sounds by breaking them into constituent parts, Rutherford's theory argued that, like the telephone, the ear responded to and transmitted all frequencies without distinction. Both are regarded as partially correct today.

53. See, especially, David M. Fryer and John C. Marshall, "The Motives of Jacques de Vaucanson," *Technology and Culture* 20 (1979): 257–69, and Silvio Bedini, "The Role of Automata in the History of Technology," *Technology and Culture* 5 (1965): 24–42. On Kempelen, Vaucanson, and automata more generally, see André Doyon and Lucien Liagre, *Jacques Vaucanson, Mécanicien de Génie* (Paris: Presses Universitaire de France, 1966), esp. ch. 5; Wolfgang von Kempelen, *Le Méchanisme de la parole, suivi de la description d'une machine parlante* (Vienna: J. V. Degen, 1791); Derek J. de Solla Price, "Automata and the Origins of Mechanism and Mechanistic Philosophy," *Technology and Culture* 5 (1965): 9–24; and Alfred Chapuis and Edmond Droz, *Automata: A Historical and Technical Study*, trans. Alec Reid (Neuchâtel: Éditions du Griffon, 1958). On the role such models may play in the processes of proof and / or discovery in the sciences, see Georges Canguilhem, "The Role of Analogies and Models in Biological Discovery," in A. C. Crombie, ed., *Scientific Change* (New York: Basic Books, 1963): 507–20; W. D. Hackman, "Scientific Instruments: Models of Brass and Aids to Discovery," 31–65, and Peter Galison and Alexei Assmus, "Artificial Clouds, Real Particles," 225–74, in Gooding, Pinch, Schaffer, eds. *The Uses of Experiment.*

54. Marey assumed much the same thing, hence the importance he placed on the resynthesis of movement throughout his career. As Dagognet notes, "Imitating the

movements of life remained essential. Marey indeed considered work on synthesis— that is, artificial reproduction—to be as indispensable as that on analysis and data collection" (*Étienne-Jules Marey*, 54).

55. Galison and Assmus, "Artificial Clouds, Real Particles," 225–74.

56. For a highly influential argument for divinity as ascertained from the structure of the eye, see William Paley, *Natural Theology*, vol. 1 (New York: Harper, 1870). Paley's arguments were at the core of Cambridge's curriculum for decades, and students were required not only to memorize them but to produce replicas in their own essays. D. L. LeMahieu, *The Mind of William Paley* (Lincoln: University of Nebraska Press, 1976), 153–83.

57. Edouard-Léon Scott, "Phonautographe et fixation graphique de la voix," 315, uses the phrase "Artiste sublime" to justify his own morphological approach. In fact, the influential metaphor of "God the clockmaker" alludes directly to automata, whose movements were usually driven by the most advanced developments in clockwork. See, especially, Derek J. de Solla Price, "Automata and the Origins of Mechanism and Mechanistic Philosophy."

58. Mary D. Waller, *Chladni Figures: A Study in Symmetry* (London: D. Bell, 1961); Walter Benjamin, *The Origin of the German Tragic Drama*, trans. John Osborne (London: New Left Books, 1977), 213–15; Theodor W. Adorno, "The Form of the Phonograph Record," trans. Thomas Y. Levin, *October* 55 (Winter 1991): 60–61.

59. Edison, Caveat 110, October 8, 1888, filed October 17, 1888, patent records, NjWOE.

60. Related locutions include Edison's own "The Phonograph and Its Future" and "The Perfected Phonograph," Philip G. Hubert's comparison of the evidential status of written contracts vs. phonographic ones where "the speaker's voice, inflection, accent, were so reproduced that witnesses could swear to the personality," the consequent lack of dispute over wills, "when the voice of the dead man [is] heard," or the musical recording "with every shade of expression, with all of its infinite changes over time" ("The New Talking Machines," *Atlantic Monthly* [February 1889], 262), and Edouard-Léon Scott's comments on the "continuous" record of an acoustic event, and the "graphic reproduction of all the intimate particularities," of a voice ("Sur le phonautographie: instrument propre à l'enregistration des sons, simples et composées," *Cosmos* 15 [1859]: 678, 679).

61. The preface to Samuel Johnson's *Dictionary* deploys the terms in a representative and illuminating way, as a means for distinguishing between the worthwhile and the worthless: "Nor are all the words which are not found in the vocabulary to be lamented as omissions. Of the laborious and mercantile part of the people the diction is in great measure casual and mutable: many of their terms are formed for some temporary or local convenience and, though current at certain times and places, are in others utterly unknown. This fugitive cant, which is always in a state of increase or decay, cannot be regarded as any part of the durable materials of a language, and therefore must be suffered to perish with other things unworthy of preservation." Samuel Johnson, "Preface to the Dictionary," in *Samuel Johnson: Selected Writings*, ed.

with an introduction by R. T. Davies (Evanston, Ill.: Northwestern University Press, 1965), 162.

62. Dagognet, *Étienne-Jules Marey*, 15–63.

63. Marta Braun, *Picturing Time: The Work of Etienne-Jules Marey*, (Chicago and London: University of Chicago Press, 1992), 17, 40–41; Dagognet, *Étienne-Jules Marey*, 16–18, 23.

64. Lorraine Daston and Peter Galison, "The Image of Objectivity," *Representations* 40 (Fall 1992): 81–128.

65. Much of the "basic research" in acoustics derived explicitly from musical concerns, and the categories of "harmony," "consonance," dissonance," etc. were broadly applied to all sonic phenomena. Indeed, musical tones served as the paradigm case for much of early acoustics.

66. John H. Pepper, *Pneumatics and Acoustics: Science Simplified* (London: Frederick Warne; New York: Scribner, Welford, and Armstrong, n.d. [after 1866; probably c. 1888]); emphasis added.

67. William Anderson, "An Outline of the History of Art in Its Relation to Medical Science," introductory address delivered at the Medical and Physical Society of St. Thomas's Hospital, 1885, *Saint Thomas's Hospital Reports* 15 (1886): 151–81, 170, cited in Daston and Galison, "The Image of Objectivity," 100.

68. Mary Ann Doane, "Temporality, Storage, Legibility: Freud, Marey, and the Cinema," *Critical Inquiry* 22.2 (Winter 1996): 313–43.

69. Largely because there had been no aesthetic of sound "recording" before the phonograph and therefore no preexisting criteria of creativity or genius, and because the phonograph was discussed and marketed as a science and business aid, questions of sonic aesthetics apparently never occurred to early audiences. Sound aesthetics as a more or less consciously recognized concern seems to be a product of the 1920s.

70. Max Hilzberg, "Mechanical Aids to Artistic Photography," *American Amateur Photographer* 7.5 (May 1895): 191–98 (192).

71. Leon van Loo, "Photography: What Are Its Possibilities in the Art Field Compared with Those of Painting?" *American Amateur Photographer* 7.6 (June 1895): 269.

72. Elizabeth Flint Wade, "Artistic Pictures: Suggestions How to Get Them," *American Amateur Photographer* 5.10 (October 1893): 441.

73. Francisco de Hollonda, *Four Dialogs on Art*, trans. Aubrey F. G. Bell (London: Oxford University Press: 1928), 15–16, cited in Svetlana Alpers, *The Art of Describing: Dutch Art in the Seventeenth Century* (Chicago: University of Chicago Press, 1983), 19–22 (emphasis added).

74. John Bull, Jr., "Photography in America, as Viewed by an Englishman," *American Amateur Photographer* 7.4 (April 1894): 144.

75. It is worth noting as well that visitors to a contemporary photography exhibition were overheard to comment that the images were "not at all like photographs" because they lacked the shiny surface typical of albumen prints and the "painful detail" of the overly sharp and undercomposed snapshot. See Hilzberg, "Mechanical Aids to Artistic Photography."

76. On the history of the idea of the "chance" image, see H. W. Janson, "The 'Image Made by Chance' in Renaissance Thought," *De Artibus Upuscula XL: Essays in Honor of Erwin Panofsky*, ed. Millard Meiss (New York: New York University Press, 1961): 254–66. See, more directly, Siegfried Kracauer, "Photography," *The Mass Ornament: Weimar Essays*, trans. and ed. Thomas Y. Levin (Cambridge, Mass., and London: Harvard University Press, 1995), 47–63.

77. *Literary Gazette* (London), July 13, 1839.

78. Oliver Wendell Holmes, "The Stereoscope and the Stereograph," *Atlantic Monthly* (June 1859): 744

79. *London Globe*, August 23, 1839.

80 Francis Frith, "The Art of Photography," *Art Journal*, 1859 (emphasis added).

81. Holmes, "The Stereoscope and the Stereograph," 745.

82. Elizabeth Flint Wade's "Artistic Pictures: Suggestions How to Get Them" offers a typical critique of such lack of discrimination (441).

83. As Baudelaire famously noted about this era, "Modernity is the momentary, the fleeting, the contingent."

84. Jacques Aumont, *L'oeil Interminable: peinture et cinéma* (Paris: Éditions Seguier, 1989): 37–72. It is important to note that in contrast to Alpers' view of the classical tradition as necessarily double, Aumont treats the tradition as single. In addition, Anne Friedberg has used this phrase in her book *Window Shopping: Cinema and the Postmodern* (Berkeley: University of California Press, 1993), esp. ch. 1, "The Mobilized and Virtual Gaze in Modernity." My sense of its importance includes both Aumont's meaning and Friedberg's more general application of the term.

85 Peter Galassi, *Before Photography: Painting and the Invention of Photography* (New York: Museum of Modern Art, 1981), 25. Certain of Galassi's claims in this essay seem highly problematic, especially his concern with photography's "legitimacy," but these need not impinge on the present discussion.

86. Alpers, 52.

87. See Malcolm Andrews, *The Search for the Picturesque* (Stanford, Calif.: Stanford University Press, 1989): 39–66 and 67–73. See also Deborah Jean Warner, "The Landscape Mirror and Glass," *Antiques* 105 (1974): 158–59.

88. Andrews, *Search for the Picturesque*, 68.

89. Cited in Thomas Southall, "White Mountains Stereographs and the Development of a Collective Vision," in Edward Earle, ed., *Points of View: The Stereograph in American Culture* (Rochester, N.Y.: Visual Studies Workshop, 1979): 98.

90. Emerson, "Nature," cited in Richard Rudisill, *Mirror Image: The Influence of the Daguerreotype on American Society* (Albuquerque: University of New Mexico Press, 1971), 17.

91. Wade, "Artistic Pictures: Suggestions How to Make Them," 441.

92. Wade, "Artistic Pictures," 444.

93. Ibid., 441.

94. See, for instance, Anson Rabinbach, *The Human Motor: Energy, Fatigue, and the Origins of Modernity* (Berkeley: University of California Press, 1990).

95. For an account of photographs as "transparent," and as offering an experience of mediated seeing, like that offered by one's glasses or a telescope, see Kendall Walton, "Transparent Pictures: On the Nature of Photographic Realism," *Critical Inquiry* 11.2 (December 1984): 246–77. See, as well, André Bazin's claim that film puts us " 'in the presence of' the actor. It does so in the same way as a mirror—one must agree that the mirror relays the presence of the person reflected in it—but it is a mirror with a delayed reflection, the tin foil of which retains the image." Bazin, "Theater and Cinema—Part Two," *What Is Cinema?*, trans. Hugh Gray, vol. 1 (Berkeley: University of California Press, 1967), 97–98,

96. The term "morality of objectivity" comes from Daston and Galison, "The Image of Objectivity."

97. On the Internet, see OED Entry Search for *phonograph*, under sec. 1.3b for "talking phonograph" (see "1884: *Pall Mall G.*" and "1890: R. Boldrewood *Miner's Right* [1899]"). Also on the Internet, see OED Entry Search for *photograph*, under sec. 2.2 (see "1862: Lady Morgan *Mem.* 1:21" and "1865: Bushnell *Vicar. Sacr.* iii. v. 296").

98. Galassi, *Before Photography*, 25.

2. Performance, Inscription, Diegesis: The Technological Transformation of Representational Causality

1. See Jean-Louis Baudry, "Ideological Effects of the Basic Cinematic Apparatus," 286–98, and "The Apparatus," 299–318; Jean-Louis Comolli, "Technique and Ideology" (parts 3 and 4), 421–43; and Philip Rosen, "Introduction" (to section on apparatus theory), 281–85—all in Philip Rosen, ed., *Narrative, Apparatus, Ideology: A Film Theory Reader* (New York: Columbia University Press, 1985). See also Stephen Heath, "The Cinematic Apparatus: Technology as Historical and Cultural Form," in Stephen Heath and Teresa DeLauretis, eds., *The Cinematic Apparatus* (New York: St. Martin's, 1980), 1–13.

2. Gaston Bachelard, *Les Intuitions Atomistiques* (Paris: Presses Universitaires de France, 1933); *La Formation de l'Esprit Scientifique* (Paris: Presses Universitaires de France, 1938); *L'Activité Rationaliste de Physique Contemporaine* (Paris: Presses Universitaires de France, 1951). See also W. D. Hackman, "Scientific Instruments: Models of Brass and Aids to Discovery," in David Gooding, Trevor Pinch, Simon Schaffer, eds., *The Uses of Experiment: Studies in the Natural Sciences* (Cambridge: Cambridge University Press, 1989), 31–66, and Simon Schaffer, "Newton's Prisms and the Uses of Experiment," 67–104, in the same volume. The expression "ideas made brass" dates from the late eighteenth century.

3. Patrick Maynard, "Talbot's Technologies: Photographic Depiction, Detection, and Reproduction," *Journal of Aesthetics and Art Criticism* 47.3 (Summer 1989): 263–76. While I make use of Maynard's quite important and insightful distinction between three basic technologies, he deploys them in the service of a theory of photographic depiction that does not necessarily align with the concerns of this chapter.

Nevertheless, Maynard's essay and another entitled "Drawing and Shooting: Causality in Depiction, *JAAC* 44.4 (Fall 1985): 115–29, remain two of the most underestimated essays on photography in the last fifteen years.

4. The laboratory, for example, calls upon the photograph's capacity to retain indexical traces of phenomena more forcefully than its ability to depict fictional narratives in a compelling manner. Conversely, an image's indexical properties may be of absolutely no practical consequence to a photographic fabulist, while its capacity to structure an iconically transparent visual display may appear essential.

5. Tom Gunning and I have independently searched many Parisian newspapers looking for the source of this anecdote and have found no evidence for its having actually occurred. Charles Musser has mentioned a number of references to people screaming at traveling exhibitor Lyman Howe's early screenings (Musser, personal communication).

6. Tom Gunning, "An Aesthetic of Astonishment," *Art & Text* 34 (Spring 1989): 31–45.

7. The most detailed, but still brief, discussion of such concerns is W. J. T. Mitchell's "Looking at Animals Looking," *Picture Theory* (Chicago: University of Chicago Press, 1994), which suggestively links the image of duped humans and animals to questions of social and political power. In a related fashion, Michael Taussig overlays a discussion of mimesis with questions about the relationship of the West to its supposed "primitive" others. Taussig, *Mimesis and Alterity: A Particular History of the Senses* (New York: Routledge, 1993), esp. the chapters "The Talking Machine" and "His Master's Voice."

8. Miriam Hansen's *Babel and Babylon* (Cambridge: Harvard University Press, 1991) offers the most sophisticated approach to early film speactatorship, especially as regards the public struggle over meaning. Also see Janet Staiger, *Interpreting Films* (Princeton: Princeton University Press, 1993); Judith Mayne, "Immigrants and Spectators," *Wide Angle* 5.2 (1982): 32–41; Charles Musser and Carol Nelson, *High-Class Moving Pictures: Lyman H. Howe and the Forgotten Era of Traveling Exhibition, 1880–1920* (Princeton: Princeton University Press, 1991).

9. Charles Musser, *The Emergence of Cinema: The American Screen to 1907*, vol. 1 of *History of the American Cinema*, ed. Charles Harpole (Berkeley, Los Angeles, London: University of California Press, 1990), 15–29.

10. Musser, *Emergence of Cinema*, 17–18.

11. Moreover, by concentrating exclusively on Kircher's *Ars magna lucis et umbrae* (1646), as so many historians do, we effectively suppress his interest in acoustics. In fact, Kircher's major work was *not* the *Ars magna* but his *Musurgia Universalis* (1662), which dealt, as did several other of his works, like the *Neue Hall und Thon-Kunst* (1684), with music and sound.

12. Barbara Stafford, *Artful Science* (Cambridge: MIT Press, 1994), 15.

13. Stafford, *Artful Science*, 74.

14. Ibid., 15–16.

15. Eugene Robertson, Étienne's son, visited the United States in 1825–26 and again

in 1834, giving phantasmagoria performances on both occasions. Magic lantern shows had become a familiar feature on the American landscape, especially in the Northeast. Peale's Museum, for example, included them in the 1820s. See Musser, *Emergence of Cinema*, 27, 28ff.; in addition, see George Clinton Densmore Odell, *Annals of the New York Stage*, vol. 3, *1821–1834* (New York: Columbia University Press, 1928), 165, 697, and Erik Barnouw, *The Magician and the Cinema* (New York and London: Oxford University Press, 1981), 19–22. Barnouw reports that Robertson père returned in 1825 and 1834, but Odell's references to him as "the balloonist" indicate that it was his son. Barnouw helpfully notes the influence of Robertson's style of performance and careful explanation on lantern and phantasmagoria shows more generally.

16. Musser, *The Emergence of Cinema*, 18–19.

17. Neil Harris, *Humbug: The Art of P. T. Barnum* (Chicago: University of Chicago Press, 1973), 72–89. In an effort to explain why so many spectacular hoaxes could succeed in a country so technologically educated, Harris offers us a picture of a culture and social climate wherein the humbug of a P. T. Barnum could flourish alongside and even depend upon great skepticism. The oft-noted aggressive individualism and cynical skepticism characteristic of American society, coupled with a technological utopianism nourished by the vanishing temporal and spatial frontiers overcome by the spectacular success of the railway and the telegraph, fostered an attitude both questioning and credulous. "Debunkers" paid to be tricked, then paid to discover how the trick worked.

18. The most successful and influential lecturer in American cinema history was Lyman Howe, who pioneered many important narrational and sound accompaniment techniques. His advertising motto was "High-Class Moving Pictures," although his shows included phonograph performances, lectures, and music.

19. Philadelphia daguerreotypist Marcus Aurelius Root reports this (cited in Harris, *Humbug*, 68–70).

20. Examples of popular successes include Gaston Tissandier, *Scientific Recreations* (London: Ward, Lock, 1883); *Recreations in Mathematics and Natural Philosophy*, a translation by Charles Hutton of Jean-Étienne Montucla's well-known edition of Jacques Ozanam (London: Thomas Tegg, 1840); R. E. Peterson, ed., *Familiar Science; Or, The Scientific Explanation of Common Things* (Philadelphia: Sower, Barnes, and Potts, 1852).

21. For example, stereograph producer Underwood and Underwood published their "Travel System" which, like other such books, routinely included a technical description of the effect of binocularity both in normal vision and through the stereoscope. Intertwined was a spiel touting the "obvious superiority" of binocular images in general and of their own devices and pictures in particular. Albert E. Osborne, *The Stereograph and the Stereoscope: With Special Maps and Books forming a Travel System* (New York: Underwood and Underwood, 1909).

22. *Moving Picture World* (hereafter, *MPW*), June 15, 1907, 235; *MPW*, May 18, 1907, 166; *MPW*, June 8, 1907, 212; *MPW*, October 26, 1907; *MPW*, August 15, 1908, 122.

23. See, for instance, *Views and Film Index* (hereafter, *VFI*), July 20, 1907, 8; *VFI*, September 7, 1907, 3.

24. Eric Smoodin, "Attitudes of the American Print Media Toward Motion Pictures," unpublished paper, cited in Richard deCordova, "The Emergence of the Star System in America," *Wide Angle* 6.4 (1985): 5. A similar discourse concerning the altered limits of the sensible existed with regard to the phonograph. Edison, among others, noted that subsonic vibrations could be recorded and, by increasing playback speed, be brought into the audible range; analogous to the instantaneous photo, it could allow us to hear nuances previously inaccessible to the ear. See Edison, "The Perfected Phonograph," *North American Review* 146 (May 1888): 648–49.

25. See, for example, Warwick Trading Co. ads in *Cinematography and Bioscope Magazine* 2 (May 1906); "Two Stories Concerning One Man," *MPW*, April 20, 1907; William Bullock, "How Moving Pictures Are Made and Shown," *MPW*, August 10, 1907, 360; "How the Cinematographer Works and Some of His Difficulties," *MPW*, June 8, 1907, 24; and "The Moving Picture Man," *MPW*, August 24, 1908, 156–57. For a parallel sense of still photography as utterly analogous to hunting, and therefore radically dependent upon the photographer's literal position and acts of visual attention, see Valentine Blanchard, "The Procedure of the Painter and Photographer Contrasted," *Pacific Coast Photographer* 2.3 (April 1893): 274; "Letters to a Friend: Who Wishes to Make Pictures with His Camera," *Pacific Coast Photographer* 2.12 (January 1894): 473, 476; and "Composition," *Pacific Coast Photographer* 1.6 (July 1892): 104.

26. Among others, see Osborne, *The Stereograph and the Stereoscope*, 72–73. Underwood and Underwood was one of the major stereograph vendors of the nineteenth century. In a chapter on stereoscopic travel, Osborne argues for an understanding of images as a prosthetic form of experience, claiming that audiences "may gain a distinct sense or experience of being in [the depicted] places." Developing his case, he argues that "the traveler seeks experiences of places, not the places themselves," suggesting that critics "are making . . . entirely too much of the presence or absence of the real Rome," since what we seek are *"experiences of being in Rome."* Reducing the material objects and locations to *"merely the means* of inducing within us those states of our conscious selves . . . which we seek in traveling," he describes a mode of picturing and of spectator engagement based on the notion of represented and shared perception.

27. "Beware of Dog," *MPW*, July 16, 1907, 277; "Who Says Moving Pictures Are Not as Good as the Real Thing?" *MPW*, May 23, 1908, 458; "More Realism," *MPW*, December 12, 1908, 454; "Film Realism," *MPW*, November 28, 1908, 427; "Shoots at Pictures," *Nickelodeon*, March 1909, 74; "A Dog Fooled by Pictures," *Nickelodeon*, February 1909, 54; "Stupid Towser," *VFI*, November 16, 1907, 9.

28. Jacques-Louis Mandé Daguerre, "Daguerreotype," in Alan Trachtenburg, ed., *Classic Essays on Photography* (New Haven: Leete's Island Books, 1980), 11; Oliver Wendell Holmes, "The Stereoscope and the Stereograph," *Atlantic Monthly* (June 1859), 744; J. H. Fitzgibbon, "Daguerreotyping," *Western Journal and Civilian* (1851): n.p.; William Henry Fox Talbot, *The Pencil of Nature* (London, 1844).

29. Edouard-Léon Scott, "Inscription automatique des sons de l'air au moyen d'une oreille artificielle," *Comptes Rendus* 53 (1861): 108–111. See also F. Moigno,

"Phonoautographe et fixation graphique de la voix," *Cosmos* 14 (1859): 314–20, and "Sur le phonoautographie: instrument propre à l'enregistration des sons, simples et composées," *Cosmos* 15 (1859): 677–79.

30. Consider Alfred Mayer's comment that "it matters not how these vibrations impress themselves on a sheet of metallic foil," in "On Edison's Talking Machine," *Popular Science Monthly* 12 (March 1878): 719; and Scott's own *La problème de la parole s'e-crivant elle-même: La France, l'Amérique* (Paris, 1878).

31. The importance of Chladni's and other methods of picturing or representing sound for the development of modern experimental science cannot be underesti-mated. These processes, which included the manometric flame, Chladni's "dust fig-ures" (*klangfiguren*), Scott's *phonautographe*, and the later phonograph, to name a few, formed the conceptual and instrumental basis of what Étienne-Jules Marey was famously to dub the "graphical method" in the sciences.

32. P. H. Emerson, "Photography or Art," *Photographic Quarterly* (January 1892), reprinted in Nancy Newhall, *P. H. Emerson: The Fight for Photography as Fine Art* (New York: 1975), 98ff.

33. Max Madden, "Photography with a Purpose," *American Amateur Photographer* 7.10 (October 1895): 442.

34. Henry Peach Robinson's "Girl with the Poison Bottle" (1857) and "Fading Away" (1858) exemplify this type of image. See Robinson, *The Elements of a Pictorial Photograph* (New York: Arno, 1973).

35. Despite Peirce's occasional assertions that icon, index, and symbol are three basic types of signs, in his own logical system they might better and more clearly be given a different status. We might argue that rather than signs, these three classfica-tions name what Peirce would call "grounds." An index, then, would be a logical ground upon which a representamen (or sign) and an object mutually determine an interpretant. Moreover, the notion of ground reintroduces the idea that certain logi-cal relations must be learned, hence, semiotic presumes a crucially social dimension. Peirce's work is crucial for an understanding of the new media in the nineteenth cen-tury, but is still too frequently misunderstood.

36. André Gaudreault, *Du littéraire au filmique: système du récit* (Paris: Klincksieck, 1988).

37. Marta Braun, *Picturing Time: The Work of Etienne-Jules Marey*, (Chicago and London: University of Chicago Press, 1992), 75–85, esp. 75, 79–83.

38. Early music recorders developed special instruments (violins with mega-phones attached to their bridges, etc.) designed to concentrate their sound for record-ing, or rearranged performers to balance recorded volumes, developed new record-ing techniques and devices and, finally, new systems for reproduction, all with the explicit goal of reproducing the effects associated with live performance.

39. Marey's *costumes* affected neither the movement of his subjects nor the objec-tivity of the inscriptional process. They did, however, direct the overall process toward certain representational ends.

40. Braun, *Picturing Time*, 81.

41. Lorraine Daston and Peter Galison, "The Image of Objectivity," *Representations* 40 (Fall 1992): 81–128. Alfred M. Mayer's book, *Sound: A Series of Simple, Entertaining, and Inexpensive Experiments in the Phenomena of Sound for the use of Students of Every Age* (London: Macmillan, 1891), in fact, organizes its course of experiments in such a way that students are explicitly taught the causal conditions under which, for example, a sinusoidal wave may be understood as tracing (and representing) a vibratory movement. The sequence of experiments is designed to teach students not only why such records may be understood as indexical but also the more general skill of how inferentially to link a particular formal feature (sinusoidal traces) to particular causal circumstances of image production (direct tracing of a vibrating stylus), and ultimately to a specific phenomenon (periodic, back-and-forth vibratory movement). Other attempts to minimize mediation concentrated on regularizing the practices of representation. Similarly, an 1893 *New York Daily Tribune* article, "Photography Proved the Signature," describes a photograph's pivotal role in determining the status of a supposedly erased signature on a deed. "The photography was done *in the presence of* the court clerk who *refused to let the deed go out of his sight.* The negative revealed traces of the missing signature . . . the court pronounced the evidence conclusive." The simultaneous presence of both clerk and document, coupled with the sworn authenticity of the experience (one assumes the clerk signed some document) are enough to guarantee the validity of the photographic document. The clerk is obviously present to ensure that the photographer operates in good faith. The camera or the photo itself could not guarantee authenticity, but strictly regulated codes of practice might. *New York Daily Tribune,* November 26, 1893, 21.

42. See, for example, *New York Daily Tribune,* July 7, 1896, 6; December 28, 1897 (sup.), 4; July 4, 1897 (sup.), 16. All refer to this "old saying" in order to question it.

43. For an illuminating discussion of the evdiential status of the X ray, especially in malpractice suits, see Daston and Galison, "The Image of Objectivity," 104–114.

44. Further examples are the novel *Captain Kodak* by Alexander Black, the short story "A Strange Witness," and the article "Saved by a Photograph," *Charterhouse Photographic Art Journal* 18 (April 1, 1889), 73–75.

45. Boucicault prepares this line with the following quotation: "You thought that no witness saw the deed, that no eye was on you; but there was . . . the eye of the Eternal was on you—the blessed sun in heaven, that, looking down, struck upon this plate the image of the deed." The equation of new technical media with divine powers of surveillance was by no means restricted to the photograph. An 1883 essay remarks, "Whether it be so or not, they are phonographed in the mind of God." See OED reference to "G. Rogers in Spurgeon Treas. Dav. Ps. cxxxix. 2–4."

46. Edison, of course, suggested exactly this possibility in his first essay on the device, "The Phonograph and Its Future," *North American Review* 126 (May-June 1878): 527–36.

47. In fact, an article from 1861 already criticizes Boucicault's device. "Really, Mr. Boucicault, you must think people are indeed ignorant of photography and its

processes, if you think they can accept [that] process as genuine." "Theatrical Photography—*The Octoroon*," *Photographic Journal* (December 16, 1861): 340.

48. "Judicial Photography," *Photographic Journal* (January 15, 1897): 107. See also J. L. Hurd, "Photographic Portraits vs. Camera Pictures," *The Photographic Eye and the Eye* (August 29, 1865).

49. I discuss this topic in more detail in my Ph.D. dissertation (University of Iowa, 1992), "Standards and Practices: Representational Technologies and the American Cinema" (ch. 2).

50. Anon., "The Photograph as False Witness," *Photographic News* (1885).

51. Ethel Egberta Thompson, "Motography in Fiction," *Nickelodeon*, February 4, 1909, 49–51. The story, "A Strange Witness," by Gilbert Parker Coleman, was published in the January 1909 issue of *Green Book Album*.

52. *MPW*, December 14, 1907, 660 (emphasis added). This formulation is reminiscent of Marey's approach. Objectivity is purely a function of inscription.

53. *Nickelodeon*, May 6, 1909, 129.

54. Howard C. Baker, "The Moving Picture Industry," *Nickelodeon*, May 6, 1909, 132.

55. "The Moving Picture Industry," *Nickelodeon*, May 6, 1909, 132 (emphasis added). The distinction proposed here seems strikingly like that between "use" and "mention" in analytic philosophies, and carries the same difficulties within it.

56. *MPW*, June 15, 1907, 235.

57. Ibid. (emphasis added).

58. See *MPW*, June 15, 1907, 234–35, "Moving Pictures Amuse and Instruct: Wild West Scene Made in Brooklyn," *VFI*, January 15, 1907, 4, and "Managing Camera Actors," *VFI* September 7, 1907, 7, for stories of this sort.

59. [Anon.], "A Wonderful Invention: Speech Capable of Indefinite Repetition from Automatic Records," *Scientific American* (December 22, 1877). Also see Roland Gelatt, *The Fabulous Phonograph, 1877–1977*, 2d rev. ed. (New York: Macmillan, 1977), 24–32; Oliver Read and Walter L. Welch, *From Tin Foil to Stereo: Evolution of the Phonograph* (Indianapolis, Kansas City, New York: Bobbs-Merrill, 1976), 11–24 (esp. 11–12).

60. John Harvith and Susan Edwards Harvith, *Edison, Musicians, and the Phonograph: A Century in Retrospect* (New York, Westport, Conn., and London: Greenwood Press, 1987), 12–13, 17; George L. Frow, *The Edison Disc Phonographs and the Diamond Discs: A History with Illustrations* (Sevenoaks, Kent, England: George L. Frow, 1982), 236–42. For accounts of earlier versions of this marketing idea, see *Collier's* (October 1908) and *Munsey's* (December 1914), which includes an advertisement.

61. Anna Case's memories of this campaign are included in Harvith and Harvith, *Edison, Musicians, and the Phonograph*, 44. An account of a related test involving Case can be found in Read and Welch, *From Tin Foil to Stereo*, 202–203.

62. Victor Talking Machine Company advertisement, reproduced in André Millard, *America on Record: A History of Recorded Sound* (Cambridge and New York: Cambridge University Press, 1995), 63.

63. Harvith and Harvith, *Edison, Musicians, and the Phonograph*, 12; Frow, *The Edison Disc Phonographs*, 236.

64. The Victor Company's famous "Red Seal" series of recordings of the world's most respected opera performers almost single-handedly conferred bourgeois respectability upon the medium. The Red Seal recordings' extraordinarily high prices limited their circulation but boosted sales by implying the highest quality.

65. Paul J. Morris, "Making Music More Musical," *Musician* 21.5 (May 1916): 263, 265.

66. Harvith and Harvith, *Edison, Musicians, and the Phonograph*, 13–14. Vachel Lindsay notes the performance style dictated by the machine when he laments the neccessity for dialogue to be "elaborately enunciated in unnatural tones with a stiff interval between question and answer." Lindsay, *The Art of the Moving Picture* (New York: Liveright, 1970), 223 and, more generally, 221–24.

67. Harvith and Harvith, *Edison, Musicians, and the Phonograph*, 8.

68. "Interview with Ernest L. Stevens," in Harvith and Harvith, *Edison, Musicians, and the Phonograph*, 25–35 (esp. 27, 34–35). See also Michael Chanan, *Repeated Takes: A Short History of Recording and Its Effects on Music* (London and New York: Verso, 1995), 58.

69. Gaisberg was pianist and recordist for the Berliner Gramophone Company and, later, the first notable record "producer." See Evan Eisenberg, *The Recording Angel: The Experience of Music from Aristotle to Zappa* (New York: Penguin, 1987), 147, and Jerrold Northrup Moore, *A Voice in Time: The Gramophone of Fred Gaisberg, 1873–1951* (London: Hamish Hamilton, 1976).

70. I mean the term "pro-phonographic" to recall Étienne Souriau's "profilmic." Here I want to designate a realm encompassing all manipulations of a sound that occur anterior to the processes of technological inscription. The goal of "intelligibility" was uppermost in Edison's mind when he initially sought to market the phonograph as a stenographic instrument. Sonic intelligibility also finds its counterpart in Edison's earliest films, which set performers (in Marey-like fashion) before a black backdrop in order to reduce extraneous or confusing visual cues. The Lumière films, by contrast, were praised for their wealth of contingent "background" detail.

71. Harvith and Harvith note, for example, that Edison evaluated a test recording of Caruso's as follows: "THIS TUNE WILL DO FOR THE DISC, but not so loud and distorted as this. Take it low and sweet. THE PHONOGRAPH IS NOT AN OPERA HOUSE" (*Edison, Musicians, and the Phonograph*, 13). A funny and much later version of this dilemma occurs in the film *Chuck Berry: Hail! Hail! Rock 'n' Roll* (1987), where Keith Richards tries repeatedly to convince Chuck Berry to alter the sound of his amplifier so that it will record well. Berry stubbornly insists on sticking to his *performance* standards, however they sound when recorded.

72. Eisenberg, *The Recording Angel*, 116.

73. Evan Eisenberg offers a concise and insightful analysis of Gaisberg, Legge, and Culshaw in *The Recording Angel*, 112–22. See also Moore, *A Voice in Time*, and Elisabeth Schwarzkopf, *On and Off the Record: A Memoir of Walter Legge* (London: Faber, 1982).

74. Gelatt, *The Fabulous Phonograph*, 317–18.

75. For more on Culshaw's *Das Rheingold* and *Elektra* recordings, see Eisenberg, *The Recording Angel*, 120–22, 138, and Gelatt, *The Fabulous Phonograph*, 317–18.

76. Cf. Thomas Y. Levin, "For the Record: Adorno on Music in the Age of Its Technological Reproducibility," *October* 55 (Winter 1991): 46. According to Levin, Adorno argues that recording "should also stop trying to imitate the concert hall performance and exploit instead the destructive and constructive power of montage in a didactic fashion" (46).

77. Sergei Prokofiev, *Sergei Prokofiev: Materials, Articles, Interviews* (USSR: Progress Publishers, 1978), 35, seems to be the source for most of the information on *Alexander Nevsky*'s music. Related accounts are available in Israel V. Nestyev, *Prokofiev*, trans. Florence Jonas (Stanford, Calif.: Stanford University Press, 1960), 294–95; Victor Seroff, *Prokofiev: A Soviet Tragedy* (New York: Funk and Wagnall's, 1968), 216–20; and Harlow Robinson, *Prokofiev: A Biography* (New York: Viking, 1987), 352–53. Russell Merritt, "Recharging *Alexander Nevsky*: Tracking the Eisenstein-Prokofiev War Horse," *Film Quarterly* 48.2 (Winter 1994–95): 34–47, is the best account of this process available in English.

3. Everything But the Kitchen Sync: Sound and Image Before the Talkies

1. David Bordwell, "The Introduction of Sound," in Bordwell, Janet Staiger, and Kristin Thompson, *The Classical Hollywood Cinema: Film Style and Mode of Production to 1960* (New York: Columbia University Press, 1985), 301, 302, 303.

2. Rick Altman, "The Silence of the Silents," *Musical Quarterly* 80.4 (Winter 1996): 648–719. For an analysis of the historically defined nature of "the cinema," see also his "Introduction: Sound/History," in Altman, ed., *Sound Theory/Sound Practice* (New York and London: Routledge/AFI, 1992): 113–25.

3. On "functional near-equivalence," see Altman, "Introduction: Sound/History," 113–25, esp. 122–25.

4. Their use of this concept owes a good deal to Jan Mukarovsky's *Aesthetic Function, Norm, and Value as Social Facts* (Ann Arbor: Michigan Slavic Publications, 1970).

5. Bordwell, Staiger, and Thompson, *The Classical Hollywood Cinema*, 117–27, 132–39.

6. "The Cinematograph in Rochester, NY," *Post-Express* (Rochester, N.Y.), February 6, 1897, 14, cited in George C. Pratt, *Spellbound in Darkness: A History of the Silent Film* (Greenwich, Conn.: New York Graphic Society, 1973), 17–18.

7. See Noël Burch, "Passions, Chases, a Certain Linearisation," *Life to Those Shadows* (Berkeley and Los Angeles: University of California Press, 1990), 143–61; and Thomas Elsaesser, ed., *Early Cinema: Space, Frame, Narrative* (London: BFI, 1990), 20–24.

8. Carlo Ginzburg, "Clues: The Roots of an Evidential Paradigm," *Clues, Myths, and the Historical Method* (Baltimore and London: Johns Hopkins University Press,

1989), 96–125. This transformation recalls the opposite trajectories of aesthetic images and texts in the nineteenth century. While, with the rise of connoisseurship as a method, and the emergence of the modern notion of style, the painted or drawn image came to be defined as absolutely unique and replete—genetically different from every other image—the printed text moved in the opposite direction. As the surface of the painting "thickened" in its uniqueness and the specificities of the medium came more and more to the fore, the surface of the verbal text seemed to become more and more transparent. Thus, at a certain point, it was entirely possible to duplicate Shakespeare's works *exactly*, while it became impossible, by definition, for even the most prosaic of Rembrandt's sketches to be duplicated.

9. See Rick Altman's comments on "functional near-equivalence" in "Introduction: Sound/History," 113–25, esp. 122–25.

10. Obviously key here is Tom Gunning's "The Cinema of Attractions: Early Cinema, Its Spectator, and the Avant-Garde," *Wide Angle* 8.3–4 (1985): 63–70..

11. Robert Beynon, *Musical Presentation of Motion Pictures* (New York: Schirmer, 1921), 3; L. Gardette, "Conducting the Nickelodeon Program, *Nickelodeon*, March 1909, 79; "The Nickelodeon," *Motion Picture World* (hereafter, *MPW*), May 4, 1907, 140.

12. Gardette, "Conducting the Nickelodeon Program," 79–80.

13. Q. David Bowers, *Nickelodeon Theatres and Their Music* (Vestal, N.Y.: Vestal Press, 1986), 131. Some theaters also used Wurlitzer PianOrchestras for this purpose. These automatic players came equipped with a piano-roll changer allowing thirty minutes of uninterrupted and unsupervised music.

14. Clyde Martin, "Playing the Pictures," *Film Index*, October 22, 1910, 13; Clarence E. Sinn, "Music for the Picture," *MPW*, December 20, 1913, 1396.

15. Gardette, "Conducting the Nickelodeon Program," 79.

16. *MPW*, April 27, 1907, 119 (on Paterson, N.J.); *MPW*, December 28, 1907, 702; "The Popular Nickelodeon," *MPW*, January 18, 1908, 36 (on New York City); *MPW*, December 4, 1909, 804.

17. *Nickelodeon*, March 1909, 79.

18. *MPW*, May 25, 1907, 180.

19. *Film Index*, December 18, 1909, 5. The tradition of illustrated singing stretches back at least to 1865, when a Methodist minister gained notoriety by using a set of projected slides to illustrate "Rock of Ages." Subsequently, illustrated songs figured in Ulysses S. Grant's presidential campaign in 1868.

20. "A Beautiful Song Slide," *Nickelodeon*, February 25, 1911, 209.

21. "Free Music Graft," *MPW*, April 25, 1908.

22. Charles Musser, "The Travel Genre in 1903–1904: Moving Toward Fictional Narrative," *Iris* 2.1 (1984): 47–60; reprinted in Elsaesser, ed., *Early Cinema*, 123–32.

23. Musser, "The Travel Genre," 127–28.

24. *MPW*, March 23, 1907, 41; *MPW*, April 27, 1907, 118; *MPW*, January 11, 1908, 19–20.

25. *MPW*, January 11, 1908, 119; *MPW*, March 23, 1907, 41; *MPW*, April 27, 1907, 118; *MPW*, November 13, 1909, 675; *MPW*, July 6, 1907, 280.

26. *MPW*, May 16, 1908, 431.

27. "The Popular Illustrated Lecture," *MPW*, July 6, 1907, 280.

28. Burton H. Albee, "Delivering a Lecture," *MPW*, May 23, 1908, 453.

29. "Illustrating a Lecture," *MPW*, January 11, 1908, 19; *MPW*, November 13, 1909, 675.

30. An editorial ("The Musical End," *MPW*, July 3, 1909, 7–8) put it this way: "The house may be cheerless, and the pictures and the manners of showing them indifferent, the song slides bad, the singing discordant, but if there is a constant stream of good music going on . . . the audience will forgive a lot."

31. The 1909–1912 period in the United Kingdom saw the Chronomegaphone, Appollogramaphone, Filmophone, Replicaphone, Hepworth's Vivaphone, and Warwick's Cinephone. In the United States systems included the Cameraphone (1905), Laemmle's Synchronophone (1907), Commercial Biophone, Talka phone, and Vi-T-Phone.

32. "Perfection of Phono-cinematograph," *MPW*, September 14, 1907, 435.

33. "The Truth About Talking Pictures," *MPW*, March 20, 1909, 327.

34. See, for example, Howard C. Baker, "The Motion Picture Industry," *Nickelodeon*, May 1909, 134; "Pictures That Speak," *Nickelodeon*, June 1909, 153–55; and "A Motographic Opera," *Nickelodeon*, January 21, 1911, 73–74.

35. W. Stephen Bush, "Lectures on Motion Pictures," *MPW*, August 22, 1908, 136.

36. *MPW*, May 25, 1907, 180.

37. *MPW*, May 16, 1908, 431.

38. *MPW*, July 13, 1907, 297.

39. See *MPW*, May 23, 1908, 453, and *MPW*, July 6, 1907, 280, generally, and on Howe specifically, *MPW*, March 9, 1907, 8; *MPW*, May 16, 1908, 431; *MPW*, October 2, 1909, 42; *Nickelodeon*, June 1909, 52, and *Views and Film Index* (hereafter, *VFI*), January 18, 1908, 11.

40. "Fights of Nations" ad, *MPW*, March 9, 1907, 9–10; ad for "The Persevering Lover," *MPW*, September 14, 1907, 443; "A Motographic Opera," *Nickelodeon*, January 21, 1911, 73.

41. "Dot Leedle German Band" and "America" are typical examples of songs turned into films.

42. Altman, "Silence of the Silents," 681.

43. Cited in Gillian Anderson, ed., *Music for Silent Films, 1894–1929: A Guide* (Washington, D.C.: Library of Congress, 1988), xiii.

44. Lux Graphicus, "On the Screen," *MPW*, August 21, 1909, 312.

45. Emmett Campbell Hall, "With Accompanying Noises," *MPW*, June 10, 1911, 1296, argues against effects and even music, claiming that the audience will know it's a dog without the "bow-wow."

46. H. F. Hoffman, "Drums and Traps," *MPW*, July 23, 1910, 185. See also "Sound Effects: Good, Bad, and Indifferent," *MPW*, October 2, 1909, 441, and W. Stephen Bush, "When Effects Are Unnecessary," *MPW*, September 19, 1911, 690.

47. Altman points out that, while everyone mentions the advent of the sugges-

tions, no one mentions that they were discontinued nine months later ("Silence of the Silents," 679).

48. *New York Dramatic Mirror*, October 9, 1909.

49. Bordwell, Staiger, and Thompson, *The Classical Hollywood Cinema*, 33.

50. Gunning, "The Cinema of Attractions," 63–70.

51. All contemporary references to Howe seem to stress this, as do *MPW*, July 27, 1908, 544, where sound effects "naturally add realism"; *Nickelodeon*, January 1909, 4; and *VFI*, August 17, 1907, 3.

52. "In the Talking Picture Field," *VFI*, September 12, 1908, 6.

53. "A Notable Event," *VFI*, February 13, 1909, 8.

54. *VFI*, January 18, 1908, 11.

55. *Pittsburgh Post*, July 21, 1908, 4, quoted in Charles Musser and Carol Nelson, *High-Class Moving Pictures: Lyman H. Howe and the Forgotten Era of Traveling Exhibition, 1880–1920* (Princeton: Princeton University Press, 1991), 186.

56. Altman, "Silence of the Silents," 697. See also Musser and Nelson, *High-Class Moving Pictures*, 176, 178–79, 184ff.

57. "Exhibitions with Sense," *VFI*, January 4, 1908, 11.

58. W. Stephen Bush, "Lectures on Motion Pictures," *MPW*, August 27, 1908, 137; *MPW*, March 13, 1909, 308.

59. *MPW*, May 16, 1908, 431.

60. Van C. Lee, "The Value of a Lecture," *MPW*, February 8, 1908, 93.

61. *Nickelodeon*, December 1909, 167–68.

62. Lux Graphicus, "On the Screen," *MPW*, August 21, 1909, 312.

63. "Sound Effects: Good, Bad, Indifferent," *MPW*, October 2, 1909, 441.

64. *MPW*, January 23, 1909, 86.

65. Tom Gunning, "Non-Continuity, Continuity, Discontinuity: A Theory of Genres in Early Film," *Iris* 2.1 (1984): 101–12.

66. Burch, "Passions, Chases, a Certain Linearization," 143–61. A film like *Rube and Mandy at Coney Island* (1903) uses recognizable characters in each shot to accomplish the "linearization" of what might otherwise seem to be a succession of autonomous actualities.

67. S. L. Rothapfel, "Music and Motion Pictures," *MPW*, April 16, 1910, 593.

68. Beynon (*Musical Presentation*) devotes a long chapter to "fitting" music to the picture, and calls it "synchronizing" the film, as do Edith Lang and George West in their book, *Musical Accompaniment of Moving Pictures: A Practical Guide for Pianists and Organists and an Exposition of the Principles Underlying the Musical Interpretation of Moving Pictures* (Boston: Boston Music Co., 1920).

69. In "Silence of the Silents" (679, 715n105), Altman notes that Gregg A. Frelinger's *Motion Picture Piano Music* (Lafayette, Ind.: G. A. Frelinger, 1910) may well be the first book devoted exclusively to film music. In addition to Beynon's, and Lang and West's, books, Erno Rapée's how-to contribution was *Motion Picture Moods for Pianists and Organists* (New York: G. Schirmer, 1924).

70. Lang and West, *Musical Accompaniment*, 8.

71. Bordwell, Staiger, and Thompson (*The Classical Hollywood Cinema*), in fact, never mention them.

72. Cited in Anderson, ed., *Music for Silent Films*, xiii.

73. That is, using "stormy" music to accompany a storm, or in program music, more generally.

74. Charles Merrel Berg, "An Investigation of the Motives for and Realization of Music to Accompany the American Silent Film, 1896–1927," Ph.D. diss., University of Iowa, 1976.

75. Burch, *Life to Those Shadows*, 148–57 and passim.

76. Clarence E. Sinn, "Music for the Picture," 1245.

77. *New York Dramatic Mirror*, October 9, 1909.

78. Beynon, *Musical Presentation*, 57.

79. "Exhibitions with Sense," *VFI*, January 4, 1908, 11; "Music and Films," *VFI*, May 16, 1908, 4. See also Beynon, *Musical Presentation*, 57.

80. *MPW*, May 30, 1908, 473. For a similar critique, see "The Truth About Talking Pictures," *MPW*, March 20, 1909, 327.

81. Mary Carbine, "The Finest Outside the Loop: Motion Picture Exhibition in Chicago's Black Metropolis, 1905–1928," *Camera Obscura* 23 (May 1990): 9–41. See, for a comic account, Louis Reeves Harrison, "Jackass Music," *MPW*, January 21, 1911, 124–35, and for an analysis, Tim Anderson, "Reforming 'Jackass Music': The Problematic Aesthetics of Early American Film Music Accompaniment," *Cinema Journal* 37.1 (Fall 1997): 3–22.

82. One thinks of the possibilities if Louis Armstrong had been in the orchestra when Bigger Thomas went to see *Trader Horn* (1930). See Richard Wright, *Native Son* (New York and London: Harper, 1940), 32–38.

83. Clyde Martin, "Playing the Pictures," *Film Index*, December 10, 1910, 5. Anticipating the later norm, he argues that the "musician should never stop playing through the showing of a picture."

84. Beynon, *Musical Presentation*, 57, 56.

85. Ibid., 69.

86. Lang and West, *Musical Accompaniment*, 4.

87. Beynon, *Musical Presentation*, 28.

88. Ibid., 70, 71, 99.

89. Lang and West, *Musical Accompaniment*, 14, 33–34; Beynon, *Musical Presentation*, 76, 78–80, 102. See, as well, Rick Altman, "The Silence of the Silents," 702–703.

90. The August 1, 1908, issue of *Views and Film Index* gives an account of a manager who canceled his orchestra for a week since audiences complained about the waste of time. Clearly, the orchestra did not play during films (cited in Altman, "The Silence of the Silents," 678). Gardette, "Conducting the Nickelodeon Program," 79, similarly indicates that accompaniment was far from continuous. "Playing the Pictures," *Film Index*, March 4, 1911, 11, indicates that even in the best houses, films often left three to six minutes silent (out of approximately fifteen minutes), cited in

Altman, ibid., 682. Beynon notes that "it was not expected that they should play continuously" in the early years (*Musical Presentation*, 5).

91. Lang and West, *Musical Accompaniment*, 7; Beynon, *Musical Presentation*, 106, 107.

92. Beynon, *Musical Presentation*, 102.

93. Altman argues that this skews historical studies ("Silence of the Silents," 653–54).

94. Rapée, *Motion Picture Moods* and Erno Rapée, *Encyclopedia of Music for Pictures* (New York: Belwin, 1925). See also James M. Glover, *Jimmy Glover, His Book* (London: Methuen, 1911), 7: "Musical directors travelled with books of 'agits,' i.e., agitatos, 'slows,'—that is, slow music for serious situations—'pathetics,' 'struggles,' 'hornpipes,' *andantes*—to which all adapted numbers called 'melos' any dramatic situation was possible."

95. Beynon, *Musical Presentation*, 56, 65, 85.

96. Lang and West, *Musical Accompaniment*, 3, 5.

97. Beynon, *Musical Presentation*, 53, Lang and West, *Musical Accompaniment*, 7.

98. Beynon, *Musical Presentation*, 56, 85; Lang and West, *Musical Accompaniment*, 33–35.

99. Lang and West, *Musical Accompaniment*, 35.

100. Rapée, *Encyclopedia*, 14–15.

101. "Talkies from a Writer's Angle," *American Cinematographer* (August 1928): 30.

102. See, among others, a Chronophone ad in *MPW*, September 14, 1907, 443. See also *MPW*, May 23, 1908, 457, and *MPW*, May 30, 1908, 473.

103. Lea Jacobs and Ben Brewster, *Theatre to Cinema: Stage Pictorialism and the Early Feature Film* (Oxford and New York: Oxford University Press, 1997), 157–60.

104. The image is quite clearly drawn and filmed to match the sound track and not vice versa. See Scott Curtis, "The Sound of Early Warner Bros. Cartoons," in Altman, ed., *Sound Theory/Sound Practice*, 191–203.

105. Curtis, "The Sound of Early Warner Bros. Cartoons," coins the term "isomorphic."

106. Bordwell, Staiger, and Thompson, *The Classical Hollywood Cinema*, 298–308; Alan Williams, "Historical and Theoretical Issues in the Coming of Recorded Sound to the Cinema," in Altman, ed., *Sound Theory/Sound Practice*, 126–37.

107. Altman, "Introduction: Sound/History," 113–25, esp. 122–25.

4. Sound Theory

1. Béla Balázs, *Theory of the Film: Character and Growth of a New Art* (New York: Dover, 1970); 216; Stanley Cavell, *The World Viewed* (Cambridge: Harvard University Press, 1979), 20; Jean-Louis Baudry, "The Apparatus," in Theresa Hak Kyung Cha, ed., *Apparatus* (New York: Tanam, 1980), 47; Gerald Mast, *Film/Cinema/Movie: A Theory of Experience* (New York: Harper and Row, 1977), 216; Christian Metz, "Aural Objects," *Yale French Studies* 60 (1980): 29.

2. Alan Williams, "Is Sound Recording Like a Language?" *Yale French Studies* 60 (1980): 51–66; Rick Altman, "The Material Heterogeneity of Recorded Sound," in Altman, ed., *Sound Theory/Sound Practice* (New York and London: Routledge/AFI, 1992), 15–31; Thomas Levin, "The Acoustic Dimension: Notes on Film Sound," *Screen* 25.3 (May-June 1984): 55–68.

3. Williams, "Is Sound Recording Like a Language?" 52.

4. Ibid., 53.

5. Levin makes some provocative connections to the notion of "aura" developed in Walter Benjamin, "The Work of Art in the Age of Mechanical Reproduction," *Illuminations*, trans. Harry Zohn (New York: Schocken Books, 1969): 217–52. Unfortunately, by appeals to nearness, copresence, etc. as essential attributes of sound, Levin appears to be "mourning a loss with respect to 'live' music," despite his explicit objections to the contrary ("The Acoustic Dimension," 67).

6. Levin, "The Acoustic Dimension," 62.

7. Williams, "Is Sound Recording Like a Language?" 53. Altman makes this point vividly in "The Material Heterogeneity of Recorded Sound," 19.

8. Williams, "Is Sound Recording Like a Language?" 58.

9. Levin, "The Acoustic Dimension," 62 (emphasis added). In "For the Record: Adorno on Music in the Age of Its Technological Reproducibility," *October* 55 (Winter 1991): 23–45, Thomas Y. Levin argues persuasively against his own earlier account. I focus on Levin's earlier essay because it has been so influential in recent theories of film sound. It is cited as being a decisive contribution in the first chapters of both Kaja Silverman's *The Acoustic Mirror* (Bloomington: Indiana University Press, 1988) and Amy Lawrence's *Echo and Narcissus: Women's Voices in Classical Hollywood Cinema* (Berkeley: University of California Press, 1991), 19–21.

10. Metz, "Aural Objects," 24–32. See, especially, his comments on the social dimensions of sound's "objecthood."

11. Altman, "The Material Heterogeneity of Recorded Sound," 19–20.

12. Or, in other words (per Metz), "Le perçu et le nommé" ("Aural Objects," 26–32, esp. 29. A sympathetic reader might take Metz to be saying that: Given a particular form of cinematic narrative whose role is strictly defined within a particular economic system, in order to ensure a particular form of narrative pleasure, sound recordings are constructed in order to ensure maximum legibility so that they may take their place as clearly defined elements in a narrative structure. And thus: Given these conditions, to the extent that two sounds can be recognized as deriving from the same source and hence, of performing the same function, they are interchangeable.

13. Levin, "The Acoustic Dimension," 56.

14. Altman discusses the "Reproductive Fallacy" and the "Nominalist Fallacy" in *Sound Theory/Sound Practice*, 35–45.

15. See Williams, "Is Sound Recording Like a Language?"; Altman, "The Material Heterogeneity of Recorded Sound"; and, especially, Levin, "The Acoustic Dimension," 67.

16. To reiterate, "in sound recording, as in image recording, the apparatus per-

forms a significant perceptual work *for us*. . . . It gives an implied physical perspective on image or sound source, though not the full, material context of everyday vision or hearing. . . . More than that: we accept the machine as organism, and its "attitudes" as our own." Williams, "Is Sound Recording Like a Language?" 58.

17. Curiously, film theory's emphasis on the original sound bears a strange resemblance to the so-called homunculus problem of epistemology, especially Cartesian discussions of perception. It recapitulates, for instance, Erwin Panofsky's misguided emphasis on the "curved" retinal image in his discussion of perspective as a "symbolic form" (*Perspective as Symbolic Form* [New York: Zone Books, 1991]). Of course, no one ever *sees* the retinal image—it has a purely physiological function. Richard Rorty offers an extended analysis of this epistemological dilemma in *Philosophy and the Mirror of Nature* (Princeton: Princeton University Press, 1979).

18. Theodor W. Adorno, "The Radio Symphony," in Paul Lazarsfeld and F. N. Stanton, eds., *Radio Research 1941* (New York: Columbia University Bureau of Applied Social Research, 1941), 110–39.

19. Adorno, "The Radio Symphony," 111–12 (emphasis added). Adorno's a priori hierarchy of experiences ("actual" vs. mass-produced) shapes his entire analysis, yet he leaves its specific character virtually undiscussed, focusing instead on the (more defensible) adequate/inadequate hierarchy. His later *Composing for the Film* (New York: Oxford University Press, 1947) suggests the hierarchy derives from Walter Benjamin, and the distinction between the life performance's aura and the recorded version's "degenerate" aura. See Levin, "The Acoustic Dimension," 61, 63–64, for a helpful discussion of this distinction.

20. Adorno, "The Radio Symphony," 110–26.

21. Ibid., 134. See also Theodor W. Adorno, "A Social Critique of Radio Music," *Kenyon Review* 7.2 (1945): 208–17.

22. Levin, "The Acoustic Dimension," 67.

23. See, for example, "A Social Critique of Radio Music," and Adorno, "On Popular Music," *Studies in Philosophy and Social Science* 9.1 (April 1941): 17–49. Here, Adorno's *Introduction to the Sociology of Music*, trans. E. B. Ashton (New York: Continuum, 1962) is also relevant, esp. chs. 1–3, 6–8.

24. The transformations identified in "The Radio Symphony" produce a mode of listening Adorno describes in his *Sociology of Music* and in "On the Fetish Character of Music and the Regression of Listening," in Andrew Arato and Eike Gebhardt, eds., *The Essential Frankfurt School Reader* (New York: Continuum, 1982), 270–99. In all three works he diagnoses the musical symptoms of a far broader set of transformations he associates with the dominance of the "Culture Industry."

25. Adorno, "The Radio Symphony," 116–20.

26. Quoted in Levin, "The Acoustic Dimension," 67. In fact, even the commodification of sound accomplished by recording is not accomplished by recording popular music, since in its very structure it is already "pre-consumed" or simply structured for mass consumption. See Adorno's "On the Fetish Character in Music," 270–299.

27. Adorno, "The Radio Symphony," 111.

28. Theodor Adorno, "Opera and the Long Playing Record," trans. Thomas Y. Levin, *October* 55 (Winter 1991): 62–66. See, as well, William C. Bohn, "Front Row Center and Thirty Feet Up," *High Fidelity* 4.2 (April 1954): 37, 92–94 (discusses "concert hall realism"); Abraham B. Cohen, "Reflections on Having Two Ears," *High Fidelity* 4.6 (August 1954): 28–31, 78–83 (describes the musical uses of three dimensions); and R. D. Darrell, "18 Anvils and a Thundersheet," *High Fidelity* 9.8 (August 1958): 42–43, 93 (discusses Culshaw's recording of *Das Rheingold*, maintaining that its stereo effects have been drastically overemphasized).

A classic statement of the opposed positions can be found in the debate between music critic Conrad L. Osborne, whose *"Elektra*: A Stage Work Updated? Or a New Sonic Miracle?" *High Fidelity* 18.2 (February 1968): 77–78, argues strenuously against what he calls "radio mellerdrammer" sound effects, and John Culshaw, who defends himself with as much vigor in "'The Record Producer Strikes Back," *High Fidelity* 18.10 (October 1968): 68–71. The debate continued for years. In 1977 Graham Melville-Mason wrote disparagingly of "engineering wizardry, sleight-of-ear audio assault or general 'phoney-oustics' as practiced by the followers and successors of the 'Culshaw-cult,' " *Phonographs and Gramophones* (Edinburgh: Royal Scottish Museum, 1977), 95–99; in the same volume Boris Semenoff argues for the symbiotic relationship between opera and sound recording (101–106).

29. For an important discussion of the relation between acoustics and musical form, see Michael Forsyth, *Buildings for Music: The Architect, the Musician, and the Listener from the Seventeenth Century to the Present* (Cambridge: MIT Press, 1985), 186–93, 199–208, 243, 258–59, 289; Michael Chanan, *Repeated Takes: A Short History of Recording and Its Effects on Music* (London and New York: Verso, 1995), ch. 7. On Adorno's feelings about conductors, see Joseph Horowitz, *Understanding Toscanini* (New York: Knopf, 1987), 234–37.

30. See Adorno, "On the Fetish Character of Music," the *Sociology of Music* (ch. 1), and "On Popular Music."

31. Levin, "The Acoustic Dimension," 66.

32. Of course, in most cases there is an important *causal* relationship between the "original" sound and its recording, but there is no logical or essential reason why *that* particular relationship should be the privileged one in every theoretical discussion. There is often, consequently, also a temporal hierarchy between original and copy (the original "comes first"), but an argument from temporal priority would require Levin, for instance, to acknowledge that *echoes* (by definition, a part of the original event) did just as much "violence" to the original as any recording, and thus recording (and the original) would lose any particular theoretical priority.

33. Hearing a recorded sound is just as unique an experience as hearing one "live"—a situation hopelessly complicated by the use of recorded sounds or electronic enhancement in a live performance. This question has assumed full middlebrow urgency in the past few years, since the *New York Times* has devoted many articles (and even an entire Sunday "Arts and Leisure" section) to the topic. Adorno is quite forthright about his own standards of proper audition. In his *Sociology of Music*, Adorno

states unambiguously that definitions of listening "types" rest upon "the adequacy or inadequacy of the act of listening to that which is heard. A premise is that works are objectively structured things and meaningful in themselves, things that invite analysis and can be perceived with different degrees of accuracy." The only fully appropriate mode of musical engagement—regardless of style or form—is what he terms "structural hearing." Structural hearing is, in turn, defined by his more fully elaborated ideas about musical structure. See pages 3–5, specifically, and chapters 1–3, more generally.

34. Levin argues, in fact, that Adorno's earliest and latest writings on sound recording implicitly argue for a montage aesthetic of sound against duplicating a concert hall performance ("For the Record," 46).

35. Levin, "The Acoustic Dimension," 66.

36. Forsyth (Buildings for Music) offers a detailed history of the interrelationship between musical form, architecture, and recording.

37. Altman's "The Material Heterogeneity of Recorded Sound" is an important point of reference. Very often it is precisely those characteristics of a sound which are most salient for us which indicate that we are not getting the "best" perspective on it, since they indicate, among other things, whether the sound is directed toward us, whether the source is advancing toward or retreating from us, etc.

38. Levin, "The Acoustic Dimension," 67.

39. Ibid., 66.

40. It is perhaps symptomatic that he defines "technology" more or less as an object, rather than a set of cultural and social practices. See Raymond Williams's arguments about the importance of art as practice rather than object, and his analysis of TV as a cultural form rather than as a device. Williams, Problems in Materialism and Culture: Selected Essays (London: New Left Books, 1980), and Williams, Television: Technology and Cultural Form (New York: Schocken, 1974).

41. Levin, "The Acoustic Dimension," 65.

42. Heidegger's much-discussed example of the hammer in Being and Time seems to support my argument. A hammer, so the argument goes, is only a hammer through its use, i.e., in view of hammering. Indeed, the hammer "itself" is said to "withdraw" in its use. As such, tools are always already inserted in a pragmatic context which fundamentally defines them. See, for example, Being and Time, trans. Edward Robinson and John Macquarrie (New York: Harper and Row, 1962), 95–105 and passim.

43. Martin Heidegger, "A Dialog on Language," in On the Way to Language, trans. Peter D. Hertz (San Francisco: Harper and Row, 1971), 17.

44. Veronique M. Foti, "Representation and the Image: Between Heidegger, Derrida, and Plato," Man and World 18 (1985): 65. Foti does a commendable job sorting out the positions of these three thinkers on the nature of the representational image. I am indebted to her interpretation.

45. See Altman, "The Material Heterogeneity of Recorded Sound," 15–31, esp. 29.

46. It might be argued that, ultimately, "voices" (be they Al Jolson's, Bert Lahr's, or Mischa Elman's violin) were the only sounds that Hollywood had an inherent need

to reproduce accurately in synchronization. Although the theoretical discourse promoted the importance of recording music (an emphasis derived from the Vitaphone's "disc" format?), almost every "problem" with the apparatuses involved either synchrony or dialogue—rarely if ever inaccurate rendering of instrumental timbre.

47. Although in the Vitaphone shorts (e.g., the *Tannhäuser* overture) the moving violin bow, the cymbal crash, and various other visible manifestations of sync replace the moving lips, it is not difficult to see the analogy between these types of sounds and speech.

48. Recorded musical accompaniment for films such as *Don Juan* (1926) and *Old San Francisco* (1927) was important to the extent that it further standardized the commodity, but the added cost to producers could not have been an effective inducement to convert to sound. The musical and sound effects accompaniment typical of both films was not necessarily any better synchronized than that provided by a gifted theater orchestra, so novelty was minimal. Warner stood to profit most through the presentation of vaudeville and concert acts who were salable precisely by providing their signature sounds in a manner which left no doubt that the persons depicted on the screen were producing those sounds. Concern with sound source was prominent in Bell Labs' Public Address research as well. See, e.g., T. A. Dowey, "Public-Address Systems," *Bell Laboratories Record* 3.2 (October 1926): 50–56.

49. Aside, that is, from certain forms of (narrative-driven) ethnic accent. The importance of these deviances from the King's (or rather Bell Labs') English is their ability to provide motivations for certain forms of melodramatic narrative. Warners' *Old San Francisco*, *Noah's Ark* (1929), and *The Jazz Singer* (1927) all attest to this use of accent (the first only in intertitles). Moreover, such accents rarely if ever impede dialogue intelligibility, while they often provide important character information and plot complication.

50. See Rick Altman's discussion of "for-me-ness" in "The Technology of the Voice" (Part 1), *Iris* 3.1 (1985): 3–20.

51. David Bordwell, *Narration in the Fiction Film* (Madison: University of Wisconsin Press, 1985), 9.

52. Rick Altman, "Sound Space," in Altman, ed., *Sound Theory/Sound Practice*, 46–64.

53. Interestingly, the tendency to think of sound representation as records of perception duplicates the early tendency to treat the individual shots of, say, a chase film as autonomous wholes whose unity is provided by point of view rather than a higher-level narrative unity. See Tom Gunning, "Non-Continuity, Continuity, Discontinuity: A Theory of Genres in Early Film," *Iris* 2.1 (1984): 101–12; Charles Musser, "The Travel Genre in 1903–1904: Moving Towards Fictional Narrative," *Iris* 2.1 (1984): 47–60; and André Gaudreault, "The Infringement of Copyright Laws and Its Effects (1900–1906)," *Framework* 29 (1985): 2–14.

54. J. C. Steinberg, "Effects of Distortion Upon the Recognition of Speech Sounds," *Journal of the Acoustical Society of America* (hereafter, *JASA*) 2.4 (October 1930): 132.

55. Franklin L. Hunt, "Sound Pictures: Fundamental Principles and Some Factors Which Affect Their Quality," *JASA* 2.4 (April 1931): 482.

56. For example, Hunt, "Sound Pictures," 481–82. It also makes clear that very "normal" kinds of sound recording paid no heed to the credo which claimed that the "eyes and ears" of the recording apparatuses had to be linked like those in a real body. It further indicates that at least a portion of the technicians involved in research around problems in sound film were aware that Hollywood was primarily in the business of selling narratives, not optical and acoustic effigies of real experience. For a good example of a technician grappling with the pair intelligibility/naturalness, see Carl Dreher, "Recording, Re-recording, and Editing of Sound," *Journal of the Society of Motion Picture Engineers* 16.6 (June 1931): 756–65.

57. Hunt, "Sound Pictures," 499.

58. A notable exception are those sound effects that require a strong dose of spatial acoustics in order to be generically recognizable. Hence we often get sound/image scale mismatches which do not disturb us in the least because the rendering of the sound stresses its "name" rather than its actuality.

59. Dreher, "Recording, Re-recording, and Editing of Sound," 756. Rick Altman has deftly explicated the argumentative maneuvers in this article, showing how some rhetorical sleight-of-hand enables Dreher to mediate between opposed positions. Altman, "Technology of the Voice," 15–17.

60. John L. Cass, "The Illusion of Sound and Picture," *Journal of the Society of Motion Picture Engineers* 14.3 (March 1930): 325.

61. Bordwell, *Narration in the Fiction Film*, 3–12, 99–113.

62. Rick Altman "Toward a Theory of the History of Representational Technologies," *Iris* 2.2 (1984): 111–25.

63. David Bordwell, Janet Staiger, Kristin Thompson, *The Classical Hollywood Cinema: Film Style and Mode of Production to 1960* (New York; Columbia University Press, 1985), 4–6.

64. Harvey Fletcher, *Speech and Hearing* (New York: Van Nostrand, 1929), 280–85. This remarkable work stood as the standard of acoustic research for at least two decades and merits a study of its own.

65. Joseph Maxfield and H. C. Harrison, "Methods of High Quality Recording and Reproduction of Sound Based on Telephone Research," *Bell System Technical Journal* 5.3 (July 1926): 493–523; Elisabeth Schwarzkopf, *On and Off the Record: A Memoir of Walter Legge* (London: Faber, 1982), 5, 16, 73, 143; James G. Stewart, "Personal Report," in Evan William Cameron, ed., *Sound and the Cinema: The Coming of Sound to American Film* (Pleasantville, N.Y.: Redgrave, 1980), 38–61, esp. 42–44.

66. See, among others, F. L. Hopper, "The Measurement of Reverberation Time and Its Application to Acoustic Problems in Sound Pictures," *JASA* 2.4 (April 1931): 499–505; T. A. Dowey, "Public-Address Systems," 50–56; Walter A. MacNair, "Optimum Reverberation Time for Auditoriums," *Bell System Technical Journal* 9.2 (January 1930): 242–56; and Vern O. Knudsen, "The Hearing of Speech in Auditoriums," *JASA* 1.1 (October 1929): 56–82.

67. Cf. Cohen, "Reflections on Having Two Ears," 28–31. For a good discussion of the aesthetic implications of this distinction, see Osborne, *"Elektra*: A Stage Work Updated? Or a New Sonic Miracle?"* 77–78, and an opposing viewpoint, Culshaw, "The Record Producer Strikes Back," 68–71.

68. David Bordwell and Kristin Thompson, *Film Art: An Introduction*, 3d ed. (New York: McGraw-Hill, 1990), ch. 8.

69. Mary Ann Doane, "The Voice in Cinema: The Articulation of Body and Space," *Yale French Studies* 60 (1980): 33–50.

70. In other words, the meaning of any writing is radically dependent upon context and therefore cannot be understood as self-defining, or as defined by any *particular* discursive situation. This form of iterability is constitutive speech as well, which, I would argue, is just as "inscribed" in particular contexts as is writing.

71. See, for example, Edwin Arthur Krows, "Sound and Speech in the Silent Film," *American Cinematographer* 12.1 (May 1931): 14–15, 26; and John L. Cass, "The Illusion of Sound and Picture," 323.

72. Altman, "Technology of the Voice," 3–20.

73. Hunt, "Sound Pictures," 482.

74. J. L. Austin, *How to Do Things with Words* (Cambridge: Harvard University Press, 1962). My argument here borrows a great deal from Derrida's (in)famous "Signature Event Context," in *Margins of Philosophy*, trans. Alan Bass (Chicago: University of Chicago Press, 1982).

75. See Jacques Derrida, *Edmund Husserl's "Origin of Geometry": An Introduction* (Lincoln: University of Nebraska Press, 1989).

5. Standards and Practices: Aesthetic Norm and Technological Innovation in the American Cinema

1. David Bordwell, Janet Staiger, Kristin Thompson, *The Classical Hollywood Cinema: Film Style and Mode of Production to 1960* (New York: Columbia University Press, 1985).

2. Pierre Bourdieu, *Outline of a Theory of Practice*, trans. Richard Nice (Cambridge and New York: Cambridge University Press, 1977), 78.

3. Bourdieu, *Outline of a Theory*, 79.

4. David Bordwell, "The Introduction of Sound," in Bordwell, Staiger, and Thompson, *The Classical Hollywood Cinema*, 298–307; Douglas Gomery, "The Coming of Sound to the American Film Industry: A History of the Transformation of an Industry," Ph.D. diss., University of Wisconsin, Madison, 1975; Alan Williams, "Historical and Theoretical Issues in the Coming of Recorded Sound to the Cinema," in Rick Altman, ed., *Sound Theory/Sound Practice* (New York and London: Routledge/AFI, 1992), 126–37.

5. David Bordwell and Janet Staiger, "Technology, Style, and Mode of Production," in Bordwell, Staiger, and Thompson, *The Classical Hollywood Cinema*, 258.

6. On this point, see David Noble, *America by Design* (New York: Knopf, 1977).

7. I refer most specifically to the tradition of apparatus theory, which sought to develop a critique of the ideological work embodied in and performed by the basic apparatuses of film. Most conspicuously, this involved a critique of the notion of the bourgeois subject *necessarily* implied, and in fact produced, by the use of lenses ground to create "Renaissance perspective." While doubtless there are strong connections between the rise of this subject and the form and ideology of Hollywood films, this critique seems overly reductive and ultimately ahistorical. For a good and sympathetic introduction, see Philip Rosen, ed., *Narrative, Apparatus, Ideology: A Film Theory Reader* (New York: Columbia University Press, 1986), 281–85, 286–372, and generally, the editor's introductions to the issues addressed by the text as a whole.

8. Joseph P. Maxfield and H. C. Harrison, "Methods of High Quality Recording and Reproducing Based on Telephone Research," *Bell System Technical Journal* 5.3 (July 1926): 494–95.

9. Harry F. Olson and Frank Massa, "On the Realistic Reproduction of Sound with Particular Reference to Sound Motion Pictures," *Journal of the Society of Motion Picture Engineers* (hereafter, *JSMPE*) 23.2 (August 1934): 64–65.

10. Joseph P. Maxfield, "Acoustic Control of Recording for Talking Motion Pictures," *JSMPE* 14.1 (January 1930): 85.

11. John L. Cass, "The Illusion of Sound and Picture," *JSMPE* 14.3 (March 1930): 325.

12. Carl Dreher, "Recording, Re-recording, and Editing of Sound," *JSMPE* 16.6 (June 1931): 757.

13. Dreher, in fact, offers this example in an article entitled "This Matter of Volume Control," *Motion Picture Projectionist* (February 1929): 11. Although written earlier than the article in the preceding note, he here expresses more concern for the continuity of the sound take as a whole, complaining about abrupt volume changes between shots.

14. Elmer C. Richardson, "A Microphone Boom," *JSMPE* 15.1 (July 1930): 41–45.

15. Wesley C. Miller, "Sound Pictures: The Successful Production of Illusion," *American Cinematographer* 10.9 (December 1929): 5–6.

16. Bourdieu, *Outline of a Theory*, 77. Bourdieu is not talking about engineers here, and perhaps states his point too emphatically. Nevertheless, as his work on norms of photography illustrates, these basic ideas are applicable to representational practice as well. See Pierre Bourdieu et al., *Un Art Moyen*, (Paris: Éditions de Minuit, 1965).

17. Maxfield and Harrison, "A Method of High Quality Recording," 498–99.

18. Maxfield and Harrison, ibid., 495 (emphasis added).

19. A term designed to recall the term *profilmic*, meaning the realm and/or event as it exists before the camera/microphone. (See also ch. 2, note 70, above.)

20. W. H. Martin and C. H. G. Gray, "Master Reference System for Telephone Transmission," *Bell System Technical Journal* 8.3 (July 1929): 536.

21. Oliver Read and Walter L. Welch, *From Tin Foil to Stereo: Evolution of the Phonograph* (Indianapolis, Kansas City, New York: Bobbs-Merrill, 1976), 373–89.

22. Frank Lawrence, "The War of the Talkies," *American Cinematographer* 9.10 (January 1929): 11.

23. L. E. Clark, "Sound Recording: Art or Trade?" *American Cinematographer* 13.2 (June 1932): 17, 32.

24. L. E. Clark, "Enter the Audiographer," *American Cinematographer* 13.3 (July 1932): 10.

25. Glenn R. Kershner, "The Age of Alibi," *American Cinematographer* 10.5.(August 1929): 11.

26. "Sound Men and Cinematographers Discuss Their Mutual Problems," *American Cinematographer* 10.5 (August 1929): 8.

27. For a reprint of the lectures delivered at the Sound School, see Lester Cowan, ed. (for the AMPAS), *Recording Sound for Motion Pictures* (New York and London: McGraw-Hill, 1931) for the first version of the class, and *Motion Picture Sound Engineering: A Series of Lectures presented to the classes enrolled in the courses in sound engineering given by the Research Council of the AMPAS, Hollywood, California* (New York: Van Nostrand, 1938) for the revamped version given in 1938. On the Academy's plans for the school, see, for example, *AMPAS Bulletin* 24 (August 15, 1929), *AMPAS Bulletin* 25 (September 25, 1929), and the Academy Library's special collections files on the Producers-Technicians group. See especially boxes 39 and 43, and the general files. On the history of the Academy, generally, see Pierre Norman Sands, *A Historical Study of the Academy of Motion Picture Arts and Sciences, 1927–1947* (New York: Arno, 1977).

28. Carl Dreher, "Stage Technique in the Talkies," *American Cinematographer* 10.9 (December 1929): 2.

29. Dreher, "Stage Technique in the Talkies," 3. According to camera reports for a range of early Vitaphone films, there is no reason to conclude that sound recording was any more troublesome than either actors or image recording. See, for example, reports from *The Maltese Falcon* (1931 version, aka *Dangerous Female*), *Hard to Handle* (1933), and *Dames* (1934), which show no real pattern of sound difficulties. Of the three, *Dames* seems to have had the most problems, in spite of its comparatively later date. Daily Production Progress Reports, Vitaphone Collection, boxes 1448A and 1448B, Warner Bros. Archives, University of Southern California.

30. See Rick Altman, "Le son contre l'image, ou la bataille des techniciens," in Alain Masson, ed., *Hollywood 1927–41* (Paris: Éditions Autrement, 1991), 74–86.

31. David Bordwell, "Space in the Classical Film," in Bordwell, Staiger, and Thompson, *The Classical Hollywood Cinema*, 53.

32. Miller. "Sound Pictures," 5–6.

33. T. E. Shea, "Recording the Sound Picture," *Bell Laboratories Record* 8.8 (April 1930): 356, 357, 358.

34. Nathan Levinson, "Re-recording, Dubbing, or Duping," *American Cinematographer* 13.11 (March 1933): 12 (emphasis added).

35. Rick Altman, "Sound Space," in Altman, ed., *Sound Theory/Sound Practice*, 54–55, makes essentially the same point.

36. Harold Lewis, "Getting Good Sound Is an Art," *American Cinematographer* 15.2 (June 1934): 65.

37. The term suggested by André Gaudreault to describe films comprised of more than one shot.

38. See Altman, "Sound Space," generally.

39. Karl Struss, "Photographing with Multiple Cameras," *JSMPE* 13.38 (May 1929): 477–78, offers a brief discussion of the process.

40. Dreher, "This Matter of Volume Control," 11.

41. The opening scene of *The Broadway Melody* (1929) in the songwriting factory seems a possible parallel. There the sound of musicians trying out new tunes appears and disappears unnaturally as the camera moves in and out of one office.

42. See Altman's "Le son contre l'image," 74–86.

43. A. Lindsley Lane, "The Camera's Omniscient Eye," *American Cinematographer* 16.3 (March 1935): 95. Effacement of cinematic technique was a well-articulated and widespread goal among technicians, and was adopted as an industrial goal. One need only to point to the assumption expressed again and again in technical journals that it was self-evident that microphones should not be seen on camera.

6. Sound Space and Classical Narrative

1. See Rick Altman's comments in "Toward a Theory of the History of Representational Technologies," *Iris* 2.2 (1984): 111–25.

2. Joseph P. Maxfield, "Technic of Recording Control for Sound Pictures," *Cinematographic Annual* 1 (1930): 412.

3. David Bordwell, Janet Staiger, and Kristin Thompson, *The Classical Hollywood Cinema: Film Style and Mode of Production to 1960* (New York: Columbia University Press, 1985), 301–302.

4. Joseph P. Maxfield, "The Vitaphone," *Bell Laboratories Record* 2.5. (July 1926): 200.

5. Maxfield, "The Vitaphone," 200.

6. Rick Altman, "Sound Space," in Altman, ed., *Sound Theory/Sound Practice* (New York and London: Routledge/AFI, 1992), 54; Joseph P. Maxfield, "Pick-Up for Sound Motion Pictures (Including Stereophonic)," *Journal of the Society of Motion Picture Engineers* (hereafter, *JSMPE*) 30.6 (June 1938): 672.

7. Altman, "Sound Space," 55.

8. H. B. Marvin, "A System of Motion Pictures with Sound," *JSMPE* 12.33 (April 1928): 86.

9. Marvin, "A System of Motion Pictures with Sound," 95.

10. Joseph P. Maxfield, "Some Physical Factors Affecting the Illusion in Sound Motion Pictures," *Journal of the Acoustical Society of America* (hereafter, *JASA*) 3.1 (July 1931): 69.

11. Maxfield, "Some Physical Factors," 70.

12. Ibid., 70.

13. T. E. Shea, "Recording the Sound Picture," *Bell Laboratories Record* 8.8 (April 1930): 357.

14. Rick Altman, "Moving Lips: Cinema as Ventriloquism," *Yale French Studies* 60 (1980): 67–79.

15. W. A. Mueller, "A Device for Automatically Controlling the Balance Between Recorded Sounds," *JSMPE* 25.1 (July 1935): 79–86 (esp. 85).

16. Edwin Arthur Krows, "Sound and Speech in the Silent Film," *American Cinematographer* 12.1 (May 1931): 14.

17. John L. Cass, "The Illusion of Sound and Picture," *JSMPE* 14.3 (March 1930): 323.

18. Cass, "The Illusion of Sound and Picture," 324.

19. Harry F. Olson and Irving Wolff, "Sound Concentrators for Microphones," *JASA* 1.3 (April 1930): 410–17; Carl Dreher, "Microphone Concentrators in Motion Picture Production," *JSMPE* 16.1 (January 1931): 23–30. Olson and Wolff placed the microphone at the base of the horn, facing it in the direction of the horn's opening, while Dreher's device placed the microphone at the focus of the parabola, facing the reflector. Both served to eliminate reflected sound and noises felt to be distracting. Dreher's model conforms more closely to current reflector/concentrators. (Further, notice the sideline microphones at any NFL game.) See also Harry Olson, "The Ribbon Microphone," *JSMPE* 16.1 (January 1931): 695–708; Harry Olson, "Mass Controlled Electro-dynamic Microphone—The Ribbon Microphone," *JASA* 3.1 (July 1931): 56–68; Julius Weinberger, Harry Olson, Frank Massa, "A Uni-Directional Ribbon Microphone," *JASA* 5.2 (October 1933): 139–47. Rick Altman offers an excellent history of microphone development in "The Technology of the Voice" (Parts 1 and 2), *Iris* 3.1 (1985): 3–20, and *Iris* 4.1 (1986): 107–20.

20. Franklin L. Hunt, "Sound Pictures: Fundamental Principles and Some Factors Which Affect Their Quality," *JASA* 2.4 (April 1931): 482. See also Elmer C. Richardson, "A Microphone Boom," *JSMPE* 15.1 (July 1930): 41–45; Bordwell, Staiger, and Thompson, *The Classical Hollywood Cinema*, 301–303.

21. See, for example, F. L. Hopper, "The Measurement of Reverberation Time and Its Application to Acoustic Problems in Sound Pictures," *JASA* 2.4 (April 1931): 499–505, and Paul E. Sabine, "The Measurement of Sound Absorption Coefficients by the Reverberation Method," *JASA* 1.1 (October 1929): 27. The January 1930 issue is a cornucopia of such studies, including V. L. Chrisler, "Measurement of Sound Transmission," *JASA* 1.2 (January 1930): 175–80; Paul E. Sabine, "Transmission of Sound by Walls," 181–201; F. R. Watson, "Coefficient of Transmission of Sound," 202–208; Wallace Waterfall, "An Audiometric Method for Measuring Sound Insulation," 209–16; Carl F. Eyring, "Reverberation Time in 'Dead' Rooms," 217–41; Walter C. MacNair, "Optimum Reverberation Time for Auditoriums," 242–56.

22. Altman, "Technology of the Voice" (Part 1), 3–20. In fact, hierarchies favoring dialogue intelligibility existed not only for different types of sounds but for different aspects of a single sound. For instance, Wesley C. Miller argues that in all acoustic

research, "We must strain every effort to keep and foster the high frequencies, perhaps even to the point of exaggerating them to a slight degree. This applies especially to dialogue." Miller, "Sound Pictures: The Successful Production of Illusion," *American Cinematographer* 10.9 (December 1929): 20–21.

23. Hunt, "Sound Pictures," 481.

24. Carl Dreher, "Recording, Re-recording, and Editing of Sound," *JSMPE* 16.6 (June 1931): 756.

25. W. H. Martin and C. H. G. Gray, "Master Reference System for Telephone Transmission," *Bell System Technical Journal* 8.3 (July 1929): 536.

26. Hunt, "Sound Pictures," 481 (emphasis added).

27. Richardson, "A Microphone Boom," 41–45.

28. Hunt, "Sound Pictures," 481.

29. Miller. "Sound Pictures," 5–6; Joseph P. Maxfield and H. C. Harrison, "Methods of High Quality Recording and Reproducing Based on Telephone Research," *Bell System Technical Journal* 5.3 (July 1926): 499–95; Martin and Gray, "Master Reference System for Telephone Transmission"; Mueller, "A Device for Automatically Controlling the Balance Between Recorded Sounds," 79–86.

30. Cass, "The Illusion of Sound and Picture," 325.

31. George Lewin, "Dubbing and Its Relation to Sound Picture Production," *JSMPE* 16.1 (January 1931): 41 (emphasis added).

32. Harry F. Olson and Frank Massa, "On the Realistic Reproduction of Sound with Particular Reference to Sound Motion Pictures," *JSMPE* 23.2 (August 1934): 67, 71.

33. Miller, "Sound Pictures," 5.

34. A. Lindsley Lane, "The Camera's Omniscient Eye," *American Cinematographer* (March 1935): 95.

35. Obviously, Warners' hopes required that these small theaters make the considerable investment of outfitting themselves for sound reproduction, which in the early years required a not inconsiderable sum.

36. David Bordwell and Janet Staiger, "Alternative Modes of Film Practice," in *The Classical Hollywood Cinema*, 378–85. Bordwell and Staiger argue that an alternative mode of film practice cannot be defined solely by particular filmic devices, but only by alternative systems—formal, narrative, and industrial. Like the classical cinema, which is defined by the stability of its characteristic systems of spatial, temporal, causal, and narrative relationships, alternative modes must exhibit different but equally coherent and systematic systems. The Vitaphone short meets these requirements, but because it may so closely resemble the classical system, its own alterity goes unrecognized.

37. Or as Michael Fried might put it, a different "structure of beholding." See, especially, *Courbet's Realism* (Chicago: University Chicago Press, 1990) and also *Absorption and Theatricality: Painting and Beholder in the Age of Diderot* (Chicago: University of Chicago Press, 1980).

38. In other words, while both modes fall "within" the classical paradigm, each sets the boundaries of practice at different points, and the "bounds of difference" as

well. See David Bordwell, "The Bounds of Difference," in Bordwell, Staiger, and Thompson, *The Classical Hollywood Cinema*, 70–84.

39. See Charles Wolfe, "The Vitaphone Shorts and *The Jazz Singer*," *Wide Angle* 12.3 (July 1990): 62.

40. Joseph Maxfield, for example defines the ideal: "Phonographic reproduction may be termed perfect when the components of the reproduced sound reaching the ears of the actual listener have the same relative intensity and phase relation as the sound reaching the ears *of an imaginary listener* to the original performance would have had." Maxfield and Harrison, "A Method of High Quality Recording," 494–95 (emphasis added). See also Olson and Massa, "On the Realistic Reproduction of Sound," 64–65; Dreher, "Recording, Re-recording, and Editing of Sound," 757; and Joseph P. Maxfield, "Acoustic Control of Recording for Talking Motion Pictures," *JSMPE* 14.1 (January 1930): 85.

41. Richard Koszarski, "On the Record: Seeing and Hearing Vitaphone," in Mary Lea Bandy, ed., *The Dawn of Sound* (New York: Museum of Modern Art, 1989).

42. As Charles Wolfe has pointed out, "direct address" to a *film* audience that we normally *oppose* to fictionality requires fictionalizing the absence of the camera, mike, and crew, so while representation is here figured as simple documentation, the process of representation is every bit as narrational as the production of a fictional diegesis. Here, however, the "character" constructed by the text as a listening agency is the spectator in the movie theater. Wolfe, "The Vitaphone Shorts and *The Jazz Singer*," 58–78.

43. K. F. Morgan, "Scoring, Synchronizing, and Re-recording Sound Pictures," *JSMPE* 13.38 (May 1929): 269–70.

44. Morgan, "Scoring, Synchronizing, and Re-recording Sound Pictures," 269–70 (emphasis added).

45. Ibid., 270–71.

46. Nathan Levinson, "Re-recording, Dubbing, or Duping," *American Cinematographer* 13.11 (March 1933): 6.

47. See André Gaudreault, *Du littéraire au filmique: système du récit* (Paris: Klincksieck, 1988); and Noël Burch, *Life to Those Shadows* (Berkeley and Los Angeles: University of California Press, 1990), 143–61.

48. Bryan Foy's *Lights of New York* (1928) and Rouben Mamoulian's *Applause* (1929), however, demonstrate that although multiple-camera shooting allows an alternation of shot scales reminiscent of classical editing, the necessary use of telephoto lenses for closer shots produces a flat-seeming space. In addition, normal positionings for shot–reverse shot dialogue scenes or for point-of-view shots is nearly impossible. In the resultant cinematic space, the camera seems to remain relatively "external" when compared to later forms of editing, in a manner comparable to TV sitcoms, as compared with TV dramas shot on film.

49. While Cary Grant and Rosalind Russell could time their dialogue to perfection in real-time performance in *His Girl Friday* (1940), they were something of an exception. The subtle nuances of the more typical film are enhanced by the constructed temporality of the edited film.

50. Of course, the term *dubbing* at this point generally meant any rerecording.

51. Morgan, "Scoring, Synchronizing, and Re-recording Sound Pictures," 268.

52. Ibid., 270.

53. Or, as Tom Gunning argues about early image editing, this manner of com-position and combination effectively suppresses one form of discontinuity, that of the material, in favor of another level of continuity, that of the narrative. Gunning, "Non-Continuity, Continuity, Discontinuity: A Theory of Genres in Early Films," *Iris* 2.1 (1984): 101–12, reprinted in Thomas Elsaesser, ed., *Early Cinema: Space, Frame, Narrative* (London: BFI, 1990): 86–94; see also André Gaudreault, "The Infringement of Copyright Laws and Its Effects (1900–1906)," *Framework* 29 (1985): 2–14.

54. Lewin, "Dubbing," 41.

55. Ibid. (emphasis added).

56. Lewin, "Dubbing," 41, 42.

57. Ibid., 43 (emphasis added).

58. Like so much else in this discussion, there are parallels here to early cinema problems of narrative. As Charles Musser has pointed out, in an effort to control the product through the point of exhibition and minimize exhibitor interference and reediting, film producers increasingly relied on narrative forms which required linear irreversibility as a condition of their legibility. See, for example, Musser, "The Nickelodeon Era Begins: Establishing the Framework for Hollywood's Mode of Production," in Elsaesser, ed., *Early Cinema*, 256–73.

59. Lewin, "Dubbing," 42 (emphasis added).

60. Levinson, "Re-recording, Dubbing, or Duping," 6, 12.

61. C. A. Tuthill, "The Art of Monitoring," *JSMPE* 13.37 (May 1929): 173.

62. Tuthill, "The Art of Monitoring," 174.

63. Ibid., 176.

64. Louis R. Loeffler, "Cutting Movietone Pictures," *American Cinematographer* 10.3 (June 1929): 19. See also Mueller, "A Device for Automatically Controlling the Balance Between Recorded Sounds," 79.

65. H. W. Anderson, "Re-recording, or Dubbing for Sound Pictures," *Projection Engineering* 2.12 (1930): np. Although we may agree that "realism" is a stylistic rather than an ontological property of films, it is nevertheless a crucial category in the daily practice of film production and exhibition—one that could be exploited as a com-modity. It is no coincidence, of course, that one of Warner Brothers' early sound suc-cesses, *Little Caesar* (1930), capitalized on this novelty with the slogan, "A murmur. A curse. A shot. A groan. Silence. Little Caesar has spoken and the whole world gasps at his desperate deeds!" *Cinema Pressbooks from the Original Studio Collections. Part One: The Press Books for United Artists 1919–49, Warner Bros. 1922–49, and Monogram Pictures 1937–46.* (Reading and Madison: Wisconsin Center for Film and Television Research: Research Publications, 1988).

66. Mueller, "A Device for Automatically Controlling the Balance Between Recorded Sounds," 79, explicitly makes this point.

67. Ibid., 79–80.

68. E. I. Sponable, "Elimination of Splice Noise in Sound-Film," *JSMPE* 26.2 (February 1936): 136–43; J. I. Crabtree and C. E. Ives, "A New Method of Blocking Out Splices in Sound Film," *JSMPE* 14.3 (March 1930): 350.

69. Lewin, "Dubbing," 42.

70. The idea of "focus" should not be taken too literally. Given the socially defined (and physiologically defined as well) nature of "aural objects," some sounds are best "recognized" in an indistinct form. As I write, for example, I hear the vague whoosh of traffic from the nearby streets, and the distant sound of voices from a schoolyard. I can identify these general sounds although I cannot *precisely* hear any of them. However, unlike a Hollywood sound track, sometimes these sounds *do* become overly "focused" and they bring themselves to my attention as the "deafening fire truck," or the "tow truck that passes through the alley every day at this time." In a film, such legibility would only be appropriate if the sound had changed its narrative function.

71. Or as Alan Williams puts it, "the signs of those sounds." Alan Williams, "Is Sound Recording Like a Language?" *Yale French Studies* 60 (1980): 58; Morgan, "Scoring, Synchronizing, and Re-recording Sound Pictures," 271–72.

72. Morgan, ibid., 283.

73. Dreher, "Sound Personnel and Organization," in AMPAS, ed., *Academy Technical Digest: Fundamentals of Sound Recording and Reproduction for Motion Pictures* (Hollywood: AMPAS, 1929–30), 85–96 (esp. 90–91).

74. L. E. Clark, "Sound Recording: Art or Trade?" *American Cinematographer* (June 1932): 32.

75. Maxfield, "Pick-Up for Sound Motion Pictures (including Stereophonic)," 672.

76. See AMPAS, ed., *Academy Technical Digest: Fundamentals of Sound Recording and Reproduction for Motion Pictures.*

77. David Bordwell, "The Introduction of Sound," in Bordwell, Staiger, and Thompson, *The Classical Hollywood Cinema*, 298–307.

78. For example, the dolly, the crane, and the mike boom were all developed in studio shops, although later developed and mass-produced by corporations like Mole-Richardson.

79. David Bordwell and Kristin Thompson, "Technological Change and Classical Film Style," in Tino Balio, ed., *Grand Design: Hollywood as a Modern Business Enterprise, 1930–1939* (Berkeley: University of California Press, 1995), 109–141, esp. 119–23.

80. To be sure, booms are not limited to or strictly defined by these systems and assumptions, but are nevertheless uniquely suited to them.

81. Dreher, "Recording, Re-recording, and Editing of Sound," 759.

82. Ibid., 759.

83. Mueller, "A Device for Automatically Controlling the Balance Between Recorded Sounds," 85.

84. See Altman's discussion of the "up-and-downer" in "Technology of the Voice" (Part 1), 3–20.

85. Mueller, "A Device for Automatically Controlling the Balance Between Recorded Sounds," 79.

86. Ibid., 85.

87. Ibid. (emphasis added).

88. Maurice Pivar, "Sound Film Editing," *American Cinematographer* 13.1 (May 1932): 12.

89. Pivar, "Sound Film Editing," 12 (emphasis added).

90. Mueller, "A Device for Automatically Controlling the Balance Between Recorded Sounds," 84.

91. Bordwell, Staiger, and Thompson as well as Mary Ann Doane and Steve Neale all make this mistake. See Altman's "Sound Space," 46–64.

92. Dreher, "Recording, Re-recording, and Editing of Sound," 757.

93. Ibid., 757, 758.

94. Ibid., 758. In the case of the crowd scene, the need to display the money invested in the construction of an actual 100-foot tower also played a part in the need for spatial literalism.

95. For a discussion of classical omniscience and restriction, see David Bordwell, *Narration in the Fiction Film* (Madison: University of Wisconsin Press, 1985), 57–59, 65–71, 126–26, 252–55, 213–17.

96. The performance of Roy Smeck in the first Vitaphone program is interesting because at certain moments he seems to be playing not to the imagined filmic audience, but to the crew on the set whom he acknowledges.

97. Dreher, "Recording, Re-recording, and Editing of Sound," 756, 757.

Conclusion

1. J. A. Norling, "The Stereoscopic Art: A Reprint," *Journal of the Society of Motion Picture and Television Engineers (JSMPTE)* 60 (March 1953): 278, 283–84, 304.

2. Raymond Spottiswoode, "Basic Principles of the Three Dimensional Film," *JSMPTE* (October 1952): 249–85.

INDEX